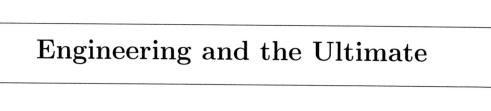

Engineering and the Ultimate

Engineering and the Ultimate:
An Interdisciplinary Investigation of Order and Design in Nature and Craft
Proceedings of the 2012 Conference on Engineering and Metaphysics

Published in the United States by Blyth Institute Press in Broken Arrow, Oklahoma

ISBN: 978-0-9752838-6-8

For author inquiries, please send email to info@blythinstitute.org. For more information, please see www.blythinstitute.org.

Information about the 2012 Conference on Engineering and Metaphysics (including videos of the original conference presentations) can be found at www.blythinstitute.org/eandm2012

1ˢᵗ printing

Blyth Institute Press

Engineering and the Ultimate

An Interdisciplinary Investigation
of Order and Design
in Nature and Craft

EDITED BY
JONATHAN BARTLETT,
DOMINIC HALSMER,
AND MARK R. HALL

Proceedings of the 2012 Conference on Engineering and Metaphysics

this book is dedicated to the search for truth in all places—both expected and unexpected

We think the question is not whether the theory of the cosmos affects matters, but whether in the long run, anything else affects them.

— G. K. Chesterton

Acknowledgements

Putting together a volume like this requires a significant amount of hard work from a large number of people. This book began as a conference, and we would like to begin by thanking Oral Roberts University for hosting the original 2012 Conference on Engineering and Metaphysics which serves as the foundation for this volume. Next we would like to thank those who provided assistance in producing the book—Heather Zeiger for providing additional copyediting, Robert Lamar, for providing design assistance, Goran Marasović, for the cover design, and Eric Holloway and Winston Ewert for providing additional help with typesetting. We would like to especially thank all of the authors in this volume for following us down an untrodden road to blaze new trails, and we would like to thank our readers for the same.

—Jonathan Bartlett, Dominic Halsmer, and Mark R. Hall

Contents

Part II   Architecture and the Ultimate   63

Chapter 4. Truth, Beauty, and the Reflection of God: John Ruskin's *Seven Lamps of Architecture* and *The Stones of Venice* as Palimpsests for Contemporary Architecture
Mark R. Hall   **65**

Part III   Software Engineering and Human Agency   97

Chapter 5. Using Turing Oracles in Cognitive Models of Problem-Solving
Jonathan Bartlett   **99**

Introduction

JONATHAN BARTLETT

The Blyth Institute

1 Philosophy and Pragmatism, Science and Engineering

When engineering comes to mind, people usually think about bridges and computers, math and science—in other words, a practical application of technology to life. Seldom do they connect engineering with philosophers and theologians. However, the truth is that these ultimate things are thoroughly embedded within engineering, even if they are rarely reflected upon. The act of construction relies on certain assumptions about the nature and limitations of the world as well as the nature and limitations of the builders. The *purpose* of a building or software program is just as important to its structure as its materials or design. These considerations are closer to philosophy or theology than they are to math and science. However, the reasoning involved in such considerations is hardly ever formalized or even made explicit.

On the flip side of the coin, engineering is often used as the ultimate test of knowledge. In America, at least, a thing is not really considered true or authentic or of value unless someone can use it to build a better product, e.g., a better phone. This concept reflects the philosophical heritage of William James, the father of the pragmatic school of philosophy. In pragmatism, the ultimate test of ideas is their "cash value"—what people can *do* with the ideas. For pragmatists, the search for *truth* for its own sake is somewhat misguided because truth cannot be apprehended until it is applied to real-world problems. For better or worse, this tends to be the philosophical framework by which most modern people live. Therefore, the ultimate test of knowledge is whether or not we can build something out of it. This pragmatic philosophy makes *testability*, the cash value of scientific ideas, so fundamental to the practice of science. It evaluates one or more scientific models based on their *practical* consequences. This is why quantum mechanics (which is testable) holds a much firmer place in science than string theory (which is not).

Such an emphasis on practicality has caused modern humanity to all but abandon speculative pursuits such as philosophy. Some recent scientists, most notably Stephen Hawking and to a lesser extent Lawrence Krauss, have openly rejected philosophy (Warman, 2011; Andersen, 2012). However, such an attitude disregards the great contributions that such speculative endeavors have had on even the most practical concerns. Modern science, in fact, is based on the rigorous and unremitting application of ideas which originated in philosophy. The conservation laws in physics are really nothing except a practical expression of the philosophical idea of *ex nihilo nihil fit* (out of nothing, nothing comes), also called the "principle of sufficient reason," which dates back to before Aristotle (Melamed & Lin, 2011).

Pragmatism itself is an outgrowth of philosophy. Although pragmatism often represents itself as being outside of speculative philosophy, its very nature arises out of speculative reasoning. Pragmatism is the outgrowth of Liebniz's concept of "the identity of indiscernibles," which was first put forth in his *Discourse on Metaphysics* (Forrest, 2012). The identity of indiscernibles states that if two things are distinct from each other, there will be some property or group of properties which will also be distinct. Pragmatism is merely the act of looking for properties that make two theories distinct from each other, or makes reality identical or distinct from some theory. If a theory is identical with reality, then the properties which hold in the theory should also hold in reality. Testing is merely probing reality to determine, on the basis of the identity of indiscernibles, whether or not the theory is identical with reality.

So, as it turns out, science, testability, and the fundamentals of physics all draw directly from speculative philosophy. Rather than science or pragmatism getting rid of the need for speculative philosophy, they prove its importance. What modern scientific thought has actually demonstrated is that sound philosophy, over time, generates rock-solid principles, principles that are so solid that most practicing scientists simply assume their truth as self-evident, often unaware that these ideas have had a history within philosophy and are the result of centuries of reflection, debate, and development. It is true that not everything in philosophy is as sound or solid as these principles. However, to reject philosophy just because newer ideas have not achieved the status as the more well-developed ones is simply to reject the general advance of knowledge because philosophers have not achieved certainty yet. Rather than dismissal, what is needed is more active engagement and development.

Not only do many modern scientists reject philosophy, but the manner in which these higher ideals are treated by the mainstream culture demonstrates the belief that everything not directly rooted in pragmatism is a mere matter of opinion. Thus, in discussions about society and social governance, any attempt to include reasonings based on the nature, purpose, and ideals of humanity, tend to be met with ramblings about how ideas such as these are private, personal matters, and not suitable for public inquiry and discourse. Rejecting such philosophical reasoning in public discourse and deeming such knowledge as a matter of personal preference is as irrational as saying, "The principle of sufficient reason doesn't apply to me." It may be true that people don't know or agree on the ideals of humanity, however, this should not be taken as an excuse to reject them as targets of public discussion, but rather a reason to explore them further and deeper.

2 Reintegrating Philosophy into Science and Engineering

As has been previously demonstrated, science's great progress has been the result of the continual and unrelenting application of sound philosophy, such as the principle of sufficient reason and the identity of indiscernibles, into all areas of inquiry. It seems reasonable, then, that additional progress can be made by explicitly recognizing the link between these fields and encouraging more cross-disciplinary dialog. While some progress in this has been made, more is desperately needed.

In 2000, Baylor University held a conference called *The Nature of Nature*. Its goal was to bring scientists, philosophers, and other academic disciplines together to talk about the ultimate nature of reality. Specifically, the question was whether *naturalism*, the idea that all of reality is a self-contained physical system, was a valid presupposition in the pursuit of science (Gordon & Dembski, 2011a). If naturalism is true, then any phenomena must be, at least in theory, describable by references to physics. Therefore, any idea not reducible to physics should not be considered a valid explanation. In such a view, for instance, *design* might be a description, but it cannot be a cause. A soul might be a useful fiction, but it cannot be a reality. Free choice is an illusion.

The *Nature of Nature* conference included the very people who built many modern scientific fields, including Francis Crick (who discovered DNA and the genetic code), Roger Penrose (whose contributions to physics are similar to Stephen Hawking's), Guillermo Gonzalez (who pioneered work on galactic habitability zones), and many other experts and professionals in science and philosophy. While the conference did not come to any particular conclusion, it was successful in moving the question concerning the ultimate nature of reality from the periphery to a more central position. Alvin Plantinga's contribution to the conference, for instance, eventually culminated in his book, *Where the Conflict Really Lies: Science, Religion, and Naturalism*, published by Oxford University Press (Plantinga, 2011b,a).

The conference eventually resulted in a book named after the conference, *The Nature of Nature* (Gordon & Dembski, 2011b). One thing, however, was markedly absent from the list of topics—any discussion of the practical consequences of any of the theories of reality offered by the conference attendees. In other words, many interesting ideas were put forth, but nothing concrete enough to result in the building of a better phone.

Since the *Nature of Nature* conference, at least two conferences which dealt with the relationship between the nature of nature, science, and engineering have been held. The first was the Royal Academy of Engineering's *Engineering and Metaphysics* seminar in 2007. The focus of this conference was the relationship between ontology (philosophy of being) and process engineering (the act of doing). In addition, a conference in 2009 titled *Parallels and Convergences* at Claremont Graduate University tackled a variety of questions focused around the large-scale goals of engineering—including space exploration and transhumanism—and their integration with the purposes of humanity.

In 2011, a conference was planned to address two major areas of integration between engineering, philosophy, and theology. The first was to examine how philosophical and theological ideas can be directly integrated into the practice of engineering. The second was to investigate how the tools of engineering can be retrofitted to analyze philosophical and theological questions. This resulted in the 2012 Conference on Engineering and Metaphysics, whose proceedings is contained in this present volume.

3   The Engineering and Metaphysics 2012 Conference

The papers in this volume, for the most part, follow the talks given in the Engineering and Metaphysics 2012 Conference.[1] Between the participants, presenters, and authors, there exists some overlap between this conference and the original *Nature of Nature* conference which inspired it. The authors come from a variety of backgrounds—academically, spiritually, and socially. Since the participants included philosophers, theologians, engineers, computer scientists, and liberal arts professors— a truly interdisciplinary group—the book has been subtitled "an interdisciplinary investigation into order and design in nature and craft."

The conference was indeed focused on investigation, an intense desire to look deeper, to search in a number of different directions as evidenced in the Table of Contents. Nature was investigated from an engineering perspective, and engineering from various perspectives on nature. Even the investigation was investigated. At the conclusion of the conference, what was produced were not finished masterpieces but new ways of holding the brush and painting the canvas, analyzing problems in engineering, philosophy, and theology through new lenses. Some ideas and approaches surely will be more successful in the long run than others, but each one benefits the discussion by looking at old questions in new and unfamiliar ways.

Some questions asked include the following: Can theological questions be asked mathematically? Can nature be rigorously analyzed in terms of purpose as well as by matter in motion? Can the human spirit be integrated into science? Can it be

---

[1]If you want to see the original talks, they are all available online at the Blyth Institute website: http://www.blythinstitute.org/eandm2012

used to analyze engineering outputs? Does it leave a distinctive mark on nature? How does theology change the goals and processes of engineering? Which of these questions are ill-posed and unworkable, and which have lasting value?

These questions are all foundational questions. This volume is not designed to be a finishing point for these types of investigations, but rather to serve as an inspiration to others to ask even better questions along the same lines. The hope is that entirely new fields will emerge at the boundaries of theology, philosophy, science, and engineering, which will ask new questions, develop new methodologies, and learn not only new answers, but also entire new ways of understanding. In short, the goal for the conference and this volume is not a final answer, but an initial inspiration. After perusing these proceedings, the reader will be challenged to reexamine the borders and boundaries of disciplines, and to think about the world in new and exciting ways.

4   Articles in this Volume

The papers in this volume are divided into four parts—Engineering, Philosophy, and Worldview; Architecture and the Ultimate; Software Engineering and Human Agency; and The Engineering of Life. Below is a short preview of each paper and its importance. While they approach very different subjects from very different perspectives, each one investigates the way in which usable knowledge can be increased by looking beyond the strict functional materialism which often dominates engineering discussions.

Reversible Universe: Implications of Affordance-based Reverse Engineering of Complex Natural Systems

This volume begins with a paper examining how scientists look at nature and suggests that reverse engineering is a fruitful methodology for natural investigations. It suggests that *purpose* is just as much of a discoverable fact of nature as is mechanism and suggests a methodology based on affordance-based reverse engineering for discovering nature's purpose as well as nature's mechanisms.

The Independence and Proper Roles of Engineering and Metaphysics in Support of an Integrated Understanding of God's Creation

The next paper analyzes the boundaries of various disciplines and shows the kinds of problems that arise from misunderstanding the proper roles and boundaries of various disciplines, including mathematics, science, engineering, and philosophy. It looks at how various spheres of knowledge do and do not interrelate, with the goal of producing a symphonic arrangement of knowledge and action.

Truth, Beauty, and the Reflection of God: John Ruskin's *Seven Lamps of Architecture* and *The Stones of Venice* as Palimpsests for Contemporary Architecture

Much of modern engineering is functional. If it works, then what more is there to do? In this paper, additional foundational considerations besides function are suggested for the practice of architecture, including moral, ethical, philosophical, and religious principles. Using John Ruskin as a plumbline, the paper provides examples of modern architecture which embody these principles and suggests ways in which these principles can be incorporated into future architectural projects.

Using Turing Oracles in Cognitive Models of Problem-Solving

Problem-solving plays a fundamental role in engineering, as one of the main tasks of an engineer is to generate creative solutions to technical problems. As such, this paper examines the question of whether humans are entirely physical or if they have a spiritual component and the impact that this has on cognitive models of problem-solving. The paper suggests Alan Turing's *oracle* concept as a way of integrating non-mechanistic human abilities into models of human insight.

Calculating Software Complexity Using the Halting Problem

Building on the previous paper, this paper gives a practical application of non-mechanistic models of problem-solving by developing a software complexity metric which is based on supra-computational abilities of humans when solving problems requiring insight. It uses the computational insolubility of the halting problem to find and measure the amount of insight required to understand a computer program.

Algorithmic Specified Complexity

The next paper in the volume considers the question of what it means for something to be engineered. Is there any property of an *engineered* system which separates it from things which are not engineered? This paper makes a technical examination of algorithmic information theory to derive a metric that the authors term *Algorithmic Specified Complexity*, which uses compressibility and context to measure the likelihood that a particular sequence is the result of intentional engineering rather than happenstance.

Complex Specified Information (CSI) Collecting

If one assumes that humans are non-mechanical and are capable of supra-computational abilities, then it may be possible to reliably harness this ability in certain applications. This paper looks at how this might be measured, tested, and harnessed programmatically. The paper includes an experimental design which, though it was not successful in this attempt, can provide a starting point for future experiments and investigations.

Developing Insights into the Design of the Simplest Self-Replicator and its Complexity

This final paper is an extended consideration of the minimal requirements for true self-replication, divided into three parts. Part 1 considers the abstract design required to allow self-replication. It analyzes what sorts of processes, components, and information is needed for any self-replication to occur. Part 2 analyzes the potential physical implementation possibilities and the various design considerations when choosing implementation materials. Part 3 compares the minimal artificial self-replicator to the self-replicators found in nature—namely cell biology. This part examines possible origin-of-life scenarios based on the analysis of the design requirements of self-replication.

References

Andersen, R. (2012). Has physics made philosophy and religion obsolete? *The At-lantic*. Available from http://www.theatlantic.com/technology/archive/2012/04/has-physics-made-philosophy-and-religion-obsolete/256203/

Forrest, P. (2012). The identity of indiscernables. In E. Zalta (Ed.), *The stanford encyclopedia of philosophy*. The Metaphysics Research Lab, winter 2012 edition. Available from http://plato.stanford.edu/archives/win2012/entries/identity-indiscernible/

Gordon, B. & Dembski, W. (2011a). The nature of nature confronted. In *The nature of nature: Examining the role of naturalism in science*. Wilmington, DE: ISI Books.

Gordon, B. & Dembski, W., Eds. (2011b). *The nature of nature: Examining the role of naturalism in science*. Wilmington, DE: ISI Books.

Melamed, Y. & Lin, M. (2011). Principle of sufficient reason. In E. Zalta (Ed.), *The stanford encyclopedia of philosophy*. The Metaphysics Research Lab, fall 2011 edition. Available from http://plato.stanford.edu/archives/fall2011/entries/sufficient-reason/

Plantinga, A. (2011a). Evolution versus naturalism. In *The nature of nature: Examining the role of naturalism in science*. Wilmington, DE: ISI Books.

Plantinga, A. (2011b). *Where the conflict really lies: Science, religion, and naturalism*. New York: Oxford University Press.

Warman, M. (2011). Stephen Hawking tells Google "philosophy is dead". *The Telegraph*. Available from http://www.telegraph.co.uk/technology/google/8520033/Stephen-Hawking-tells-Google-philosophy-is-dead.html

Part I

Engineering, Philosophy, and Worldview

This volume begins with two papers discussing the relationship between engineering and knowledge. In the first paper, Dominic Halsmer et al. argue that engineering is foundational to the search for truth as truth. It looks specifically at reverse-engineering and design recovery as a model for seeking truths about the natural world. These methodologies have been developed and used within engineering to recover a design when the specifications are not readily available, and Halsmer argues that they can be applied to the scientific endeavor to uncover the original design within nature.

In the second paper, Alexander Sich argues that engineering cannot be foundational in the search for truth because the engineering disciplines depend fundamentally on highly-focused natural scientific knowledge applied to the production of artifacts, while the particular natural sciences depend foundationally upon metaphysics for their first principles. Moreover, Sich contends that a failure to draw proper ontological distinctions between the objects studied by the natural sciences and philosophy leads to confusion over the character of inferences to the existence of the objects studied.

2

Reversible Universe: Implications of Affordance-based Reverse Engineering of Complex Natural Systems

Dominic Halsmer, Michael Gewecke, Rachelle Gewecke, Nate Roman, Tyler Todd, and Jessica Fitzgerald

Oral Roberts University

Abstract

Recent advances in the field of engineering design suggest the usefulness of the concept of affordance for reverse engineering of both man-made and natural systems. An affordance is simply what a system provides to an end-user or to another part of the system. With the current recognition that engineering concepts are playing a key role in deciphering the workings of complex natural systems such as the living cell and the human brain, affordance-based reverse engineering procedures should be considered as appropriate tools for this work. Such an approach may have important implications for philosophy and theology.

Procedures for reverse engineering and design recovery have become well-defined in several fields, especially computer software and hardware, where pattern detection and identification play important roles. These procedures can also be readily applied to complex natural systems where patterns of multiple interacting affordances facilitate the development, sustenance and education of advanced life forms such as human beings. Thinking about the human condition in terms of affordances leads to a new and fruitful interaction between the fields of science and theology, in which the field of engineering plays a key role in the dialogue. Proper understanding of the interplay between both positive and negative affordances in the context of engineering design under necessary constraints leads to a clearer worldview and a better understanding of mankind's place and purpose in the universe.

1   Introduction

A worldview consists of what one believes to be true about the universe and all of reality, including "how things work."[1] Worldviews are formed by mentally processing and filing away the accumulated experiences of life. Inaccurate worldviews can prove hazardous to one's health. Although young children do not know the inverse square law of gravitation from physics, they learn very quickly to respect the influence of gravity, or suffer the painful consequences. Other aspects of reality take somewhat longer to ascertain. But from the first moments of life, human beings enter this world as little private investigators, gathering clues as to how the world works. They are truth-seekers, especially when it comes to that which brings satisfaction and enjoyment. Upon discovery of a new object, they immediately embark on a crude type of reverse engineering exercise to determine what can be known about this object and what uses it might afford. Toddlers often go through a phase where they enjoy banging a spoon against pots and pans. Perhaps this affords them an early experience of understanding that they possess the power to control their environment, to some small degree. Or maybe they just like to make the banging sound. In either case, it results in increased hand-eye coordination and knowledge of causation.

As children get older and become more adept at reverse engineering techniques, they often pass through a phase characterized by repeated dissections of both natural and man-made objects. Perhaps taking apart complex organisms/devices facilitates the discovery of hidden connectivities, or internal affordances, which sheds light on the underlying mechanisms of operation. Or maybe they just like to see if spiders can still walk around with fewer than their original number of legs.[2] In either case, this type of behavior is often seen as a precursor to a career in science or engineering.[3] This innate

---

[1]Ken Samples gives a more technical description, writing, "A worldview forms a mental structure that organizes one's basic or ultimate beliefs. This framework supplies a comprehensive view of what a person considers real, true, rational, good, valuable, and beautiful" (Samples, 2007, p. 20). Ron Nash defines worldview as "a conceptual scheme by which we consciously or unconsciously place or fit everything we believe and by which we interpret and judge reality" (Nash, 1988, p. 24). For a thorough treatment of the concept of worldview, see David K. Naugle, *Worldview: The History of a Concept* (Naugle, 2002).

[2]Although unproductive from the point of view of the spider, this is actually an example of the "subtract and operate" reverse engineering technique for establishing component function in complex systems, discussed in Kevin N. Otto and Kristin L. Wood, *Product Design: Techniques in Reverse Engineering and New Product Development* (Otto & Wood, 2001, pp. 159–162, 204–211)

[3]This is not to suggest that budding scientists and engineers enjoy practicing cruelty to animals, but simply that they typically possess a high level of curiosity for "how things work." For a wonderful picture book to introduce children to the concepts of stewardship and sustainability issues in the

and seemingly insatiable curiosity is a very interesting feature of all human beings, especially when coupled with the extraordinary comprehensibility of the world[4], since it results in many profitable and satisfying affordances. Both humans and the external world appear to be engineered so that interactions with the environment result in vital knowledge, which plays a key role in gaining wisdom and maturity for a full and abundant life.

This paper is an investigation into the usefulness of state-of-the-art reverse engineering concepts and techniques for accurate worldview formation, with particular interest in how the field of natural theology may be influenced by thinking of nature in terms of affordances. An affordance is simply what an object provides to an "end user." In the case where the object is a more complex multi-component system, affordances are also recognized to exist internally, between interacting parts of the system. Hence, an affordance is also what one part of a system provides to another part of the system. Traditionally, reverse engineering has focused on identifying functionality, but recent engineering design research (Maier, 2008, pp. 34–37)[5] suggests that affordance-based reverse engineering may be more appropriate for handling the complexity associated with many natural systems. It may also be more helpful in design recovery, a subset of reverse engineering that attempts to work out what a system was designed to do, how it does it, and why it works the way it does. Design recovery goes beyond simply examining a system's component parts and their interactions and attempts to identify both purpose and logical organization. Although not expected to lead to definitive results in the case of natural systems, this approach contributes to a better understanding of why the universe is the way it is[6], resulting in positive contributions to the field of Christian apologetics, and consistent with a new vision for natural theology.[7]

---

animal kingdom, see Margaret Bloy Graham, *Be Nice to Spiders* (Graham, 1967).

[4]Albert Einstein famously quipped that the most incomprehensible thing about the world is that it is comprehensible. He expounded on this view by writing, "You find it strange that I consider the comprehensibility of the world (to the extent that we are authorized to speak of such a comprehensibility) as a miracle or as an eternal mystery. Well, *a priori* one should expect a chaotic world, which cannot be grasped by the mind in any way. . . . The kind of order created by Newton's theory of gravitation, for example, is wholly different. Even if man proposes the axioms of the theory, the success of such a project presupposes a high degree of ordering of the objective world, and this could not be expected *a priori*. That is the 'miracle' which is being constantly reinforced as our knowledge expands" (Einstein, 1987, p. 131).

[5]For discussion of affordance-based reverse engineering from the perspective of engineering education, see Dominic Halsmer, Nate Roman, and Tyler Todd, "Integrating the Concept of Affordance into Function-based Reverse Engineering with Application to Complex Systems" (Halsmer, Roman, & Todd, 2009b).

[6]A recent contribution in this area can be found in Hugh Ross, *Why the Universe Is the Way It Is* (Ross, 2008).

[7]An exciting vision is cast in Alister E. McGrath, *The Open Secret: A New Vision for Natural Theology* (McGrath, 2008).

2   Reverse Engineering Natural Systems

Reverse engineering is the process by which anything that has been made is analyzed to determine the original design information that went into its development.[8] Though not a major branch of engineering curriculum in academia, reverse engineering has been studied and implemented extensively in many industries where it often assists in gaining a competitive advantage. In this sense it is considered to be a mature field in practice, if not in theory. Engineering programs at some universities are now recognizing the value of reverse engineering activities as a design training exercise for students (Wu, 2008, pp. 57–59). They disassemble and analyze power tools and other man-made devices in order to see how design principles and practices guide engineers in developing a well-made product. Reverse engineering concepts are also used extensively in deciphering unfamiliar or poorly documented computer software and hardware systems (Eilam, 2005).

But recently reverse engineering techniques have proven to be surprisingly fruitful when applied to natural systems. Although the field of systems biology[9] has expanded rapidly in the past few years, there is a long history behind this approach. One of the earliest scientists/engineers to record detailed reverse engineering studies of biological systems, including extensive dissections of the human body, was Leonardo da Vinci. Concerning this work, he wrote that "the human foot is a masterpiece of engineering and a work of art." Evidently, his intimate knowledge of biological systems led him to a deep appreciation for the beautiful functionality they exhibit. In the 1600s, William Harvey discovered the detailed flow patterns of blood within the human body by asking himself how an engineer would have constructed such a system (Auffray & Noble, 2009).

In more recent times, scientists like E. O. Wilson and Daniel Dennett concur with the necessity of the reverse engineering approach. Wilson writes that "the surest way to grasp complexity in the brain, as in any other biological system, is to think of it as an engineering problem" (Wilson, 1998, p. 112). Dennett claims that "you just can't do biology without doing reverse engineering, and you can't do reverse engineering without asking what reasons there are for whatever it is you are studying. You have to ask 'why' questions" (Dennett, 1996, p. 213). Though Dennett may disagree with the idea of ultimate purpose, these are questions of a teleological nature, and other scientists and engineers are coming to the same conclusion. For example,

---

[8]Other definitions of reverse engineering emphasize the determination of specifications that allow for the reproduction of a device or system, or to provide insight into how to "reengineer" the system by incorporating updates or improvements.

[9]According to the editors of a joint issue of *IEEE Transactions on Automatic Controls* and *IEEE Transactions on Circuits and Systems: Special Issue on Systems Biology*, systems biology is "the quantitative analysis of networks of dynamically interacting biological components, with the goal of reverse engineering these networks to understand how they robustly achieve biological function" (Khammash, Tomlin, & Vidyasagar, 2008, p. 4).

Caltech researchers ask "What is/are the purpose(s) of this biological system?"[10] and suggest that biological systems be approached from an "engineer's perspective." Arthur Lander at UC Irvine has proposed a system for thinking in these terms and writes that "these elements can be seen as the foundations for a new calculus of purpose, enabling biologists to take on the much-neglected teleological side of molecular biology. 'What purpose does all this complexity serve?' may soon go from a question few biologists dare to pose, to one on everyone's lips" (Lander, 2004).

If such teleological, yet scientific questions are being asked at the micro-scale, then it seems reasonable that such questions could also be posed at the macro-scale. Furthermore, since such profitable answers are being found by reverse engineering at the level of the cell, it makes sense that good answers might also be found on a larger scale using this approach. The idea that qualitative questions can be answered through quantitative approaches is affirmed by Lander in another article where he summarizes the lessons learned from systems biology. He concludes the article with, "They teach a lesson about biology that is as important as it is surprising: sometimes, answering the most qualitative of questions – 'Why does the organism do it that way?' – succeeds only through the most quantitative of approaches" (Lander, 2007). These insights from systems biology suggest that reverse engineering of natural systems may not only reveal the inner workings of the cell, but may also assist in the acquisition of a more complete understanding of mankind's origin, place and purpose in the universe.

3 Interpreting Natural Systems

Even before the time of Christ, Greek and Roman philosophers practiced a kind of reverse engineering of natural systems, interpreting the beneficial order in the universe as an indication of a larger plan or design of a Mind (Sedley, 2009). Socrates and Plato believed that in addition to providing the initial order to the universe, this Mind also acted to sustain it at all times. About 50 years before Christ, Marcus Cicero, who brought Greek philosophy to the Romans, even suggested that various characteristics of the creating deities could be inferred from the highly ordered work of their hands. He interpreted harmonious movements in nature by referring to similarities with man-made objects when he wrote, "When we see some example of a mechanism, such as a globe or clock or some such device, do we doubt that it is the creation of some

---

[10]In Gregory T. Reeves and Scott E. Fraser, "Biological Systems from an Engineer's Point of View," the authors write that "many biologists have remarked on the apparent design of biological systems, arguing that this is a false analogy. However, evolutionary theory would predict apparent design and purpose in biological systems. Therefore, regardless of the origin of this apparent design, the analogy is, at the very least, pragmatic. Keeping this in mind we can approach a biological system from an engineer's perspective. Engineered systems were designed with a particular purpose in mind, so it would be helpful to ask, 'What is/are the purpose(s) of this biological system?' . . . Determining what [these purposes] are for a particular biological system is especially important in light of design trade-offs, and furthermore will provide clues to a systems molecular behavior" (Reeves & Fraser, 2009).

conscious intelligence? So when we see the movement of the heavenly bodies . . . how can we doubt that these too are not only the works of reason but of a reason which is perfect and divine?" (Cicero, *De Natura Deorum*, 2.38.97, trans. 1972)

Paul the Apostle wrote in a similar vein to the Romans when he penned the well known verse, "For since the creation of the world God's invisible qualities—his eternal power and divine nature—have been clearly seen, *being understood from what has been made*, so that men are without excuse" (Romans 1:20 NIV, emphasis added). Paul, being a highly educated Hebrew from Tarsus, would have been well aware of the prevailing philosophies of his audience. His words seem to mesh nicely with the idea of reverse engineering, which is about gleaning design information regarding an object, and if possible, also uncovering what may be known about the original engineer and his/her intentions. This is largely accomplished through a systematically obtained understanding of the object in the context of its surrounding culture and environment. In this verse, Paul may be referring to human understanding derived from observations of specific objects in nature, or the entire cosmos, or both.[11] It is suggested that modern reverse engineering techniques would find profitable application in both cases.

Nineteenth century natural theologians such as William Whewell and William Buckland practiced an early form of reverse engineering of natural systems, focusing on the implications for a Christian worldview. For Whewell, there was one Biblical teaching that stood out as a heuristic for science: that human beings are created in the image and likeness of God (Fuller, 2006, p. 283). This similarity between creator and creature, like the match between the complexity of the universe and mankind's ability to comprehend it, should facilitate the process of reverse engineering. Recognizing that one could never fully comprehend the transcendent engineering of the creator, nonetheless, for a Christian, there is a sense that science is the gift and privilege of "thinking God's thoughts after him."[12]

One of William Buckland's memorable expositions involved the design of Megatherium, an enormous extinct relative of the sloth (Roberts, 1999, p. 245). Leading anatomists of the time regarded this animal as having a poor and bungled design. But Buckland chose it to show, by "careful and rigorous anatomical description and then the application of reverse engineering," that it was "perfectly designed or adapted for its environment. . . . Here, for Buckland, design was not so much a scientific theory, but rather a metaphysical or theological outlook, which gave confidence or grounds for applying reverse engineering procedures"(Roberts, 1999, p. 248). It

---

[11]Recent formulations of the design argument show groups working in both directions. See Stephen C. Meyer, *Signature in the Cell: DNA and the Evidence for Intelligent Design* (Meyer, 2009) for an example of evidence from specific objects in nature, and Edward Feser, *The Last Superstition: A Refutation of the New Atheism* (Feser, 2008) for a Thomist view in which design is evident throughout the cosmos. It seems that a cumulative case for a Christian worldview, such as that described in R. Douglas Geivett, "David Hume and the Cumulative Case Argument," (Geivett, 2005) could make good use of both formulations.

[12]Often attributed to Johann Kepler. This issue is explored in Del Ratzsch, "Design: What Scientific Difference Could it Make?" (Ratzsch, 2004).

makes sense for a Christian engineer or scientist to apply such procedures in laboring under the hypothesis that the universe is an engineered system[13], without any preconceived notions about how such transcendent engineering was accomplished. This is consistent with the thinking of leading theologians of today, like Alister McGrath, who suggests that natural theology is to be understood as "the enterprise of seeing nature as creation, which both presupposes and reinforces fundamental Christian theological affirmations" (McGrath, 2006, p. 64).

McGrath also asserts that "the order of things determines how things are known . . . or, to put it more . . . formally: ontology is to be allowed to determine epistemology"(McGrath, 2006, pp. xv–xvi). Again, this is consistent with a reverse engineering mindset, which approaches the task with a certain humility, recognizing that what is discovered will in large part determine how to proceed with the overall investigation. In exploring the metaphor of nature as book, philosopher Angus Menuge writes that "a good scientific interpretation is one that allows nature to speak for itself and yet which is motivated by and connected to an overarching frame of meaning provided by revealed theology" (Menuge, 2003, p. 96).

In a response to Menuge, theoretical chemist Walter Thorson calls for maintaining a clear distinction between science and theology. He writes,

> Even the most rudimentary biosystems manifests logical organization directed to certain (limited) achievements. . . . This logical organization according to function can be explained on its own terms – as an objective aspect of a naturalistic *science*; interpretation in terms of divine agency is not essential. By such a naturalistic study of creation in its own contingent terms of reference, we would only discern the embodied logic of creaturely things themselves, not their transcendent divine purpose or design. . . . Theologically, such a situation invites the idea that God's work of creation, like his work of redemption, may be seen as the expression of a self-giving, self-emptying love: that is, creation seen as kenosis. While this view poses some difficult questions, it deserves serious consideration. (Thorson, 2003, p. 101)

But even with the attempt to keep science and theology separate, it seems likely that such theological musings will also influence aspects of an ongoing scientific approach. For those pursuing advances in both science and theology, the two fields are often found to be quite compatible, leading to many fruitful interactions.

---

[13]For explorations into the idea of the universe as an engineered system, see Dominic Halsmer et al., "The Applicability of Engineering Design Principles in Formulating a Coherent Cosmology and Worldview" (Halsmer, Halsmer, Johnson, & Wanjiku, 2008), and Dominic Halsmer et al., "The Coherence of an Engineered World" (Halsmer, Asper, Roman, & Todd, 2009a).

Figure 2.1: The Antikythera Mechanism at the National Archaeological Museum, Athens—Giovanni Dall'Orto

4   The "Artifact Hermeneutics" of Daniel Dennett

An example of a philosopher who applies reverse engineering techniques to the works of "mother nature" is found in the philosopher Dennett. He proposes that the same "artifact hermeneutics" be used when reverse engineering is applied to both man-made and biological systems (Dennett, 1990, p. 177). He further asserts that optimality considerations should be used, rather than attempting to analyze the intentions of a designer. As an example, he cites the Antikythera device, a complex geared mechanism discovered in an ancient shipwreck in 1900. Dennett contends that "it was—almost certainly—an orrery or planetarium, and the proof of that is that it would be a good orrery. That is, calculations of the periods of rotation of its wheels led to an interpretation that would have made it an accurate (Ptolemaic) representation of what was then known about the motion of the planets" (Dennett, 1990, p. 180). Dennett is correct, as far as the function of the device, but many other interesting questions might be addressed through a more complete approach to reverse engineering.

Jo Marchant details the entire story of the reverse engineering of the

Figure 2.2: Reconstruction of the Antikythera Mechanism at the National Archaeological Museum, Athens—Giovanni Dall'Orto

Antikythera device in a fascinating book (Marchant, 2009). A significant amount of design recovery was accomplished in terms of the purpose of the device and the identity and thinking of the original engineer(s). Referring to Greek letters engraved on the casing of the device, she writes, "[The reverse engineer] also noted that the letters were so precise they must have been engraved not by a labourer but by a highly trained craftsman" (Marchant, 2009, p. 55). She also recognized that the incorporation of historical and cultural information from the time period was valuable for unlocking some of the mysteries of the device. As an example, consider the following passage: "Archaeologists also studied the rest of the salvaged cargo. Their discoveries help to paint a vivid picture of when the ship sailed, where her load was being taken, and the sort of world from which she came. From there we can guess at the origins of the Antikythera mechanism itself, and how it ended up on its final journey" (Marchant, 2009, p. 61). Thus, it is clear that reverse engineering is most effective when all pertinent information is brought to bear. This is accomplished by looking at the "big picture" instead of limiting the study to the narrow set of data obtained by simply dissecting the specimen.[14]

Dennett's approach to reverse engineering through optimality considerations has recently been criticized from a couple directions. Philosopher Robert Richardson contends that such an approach to evolutionary psychology lacks the standard level of evidential support enjoyed by evolutionary biology (Richardson, 2007).[15] A second criticism, that gets more to the heart of the matter, looks specifically at Dennett's insistence on optimality as the guiding beacon of the reverse engineering enterprise. Philosophers Krist Vaesen and Melissa van Amerongen have recently published an extensive analysis of Dennett's artifact hermeneutics. They argue that "Dennett's account is implausible . . . [and] conclude that, quite in contrast to Dennett, intentional considerations play a crucial role in artifact hermeneutics, and even stronger, are necessary for the sake of simplicity and precision" (Vaesen & van Amerongen, 2008, p. 779).

Vaesen and van Amerongen claim that artifacts should be interpreted by relying both on optimality and intentional considerations, recognizing that this hampers Dennett's strategy of reverse engineering artifacts and organisms in the same way (Vaesen & van Amerongen, 2008, pp. 794–795). Their thoughts on how Dennett's unified approach might still be achieved leads to an interesting final paragraph. They write,

---

[14]For more details on this idea, see Dominic Halsmer and Jessica Fitzgerald, "Metaphysical Considerations Enhance Reverse Engineering Studies," presented at the ASA Annual Meeting, North Central College, July 29-August 1, 2011 (Halsmer & Fitzgerald, 2011), and for a discussion of the incorporation of the concept of corruption, see Dominic Halsmer and Sean McDonough, "Affordance-based Reverse Engineering of Natural Systems with Possible Corruption," Proceedings of the 2011 Christian Engineering Education Conference (CEEC), Trinity Western University, June 29-July 1 (Halsmer & McDonough, 2011).

[15]Refer to his recent book, *Evolutionary Psychology as Maladapted Psychology* (Richardson, 2007) for details of his criticisms.

Of course, it still might be possible to establish a generic interpretive program, including artifacts *and* biological items . . . What is needed to argue for the importance of intent in the interpretation of organ(ism)s, is a proof that intent reveals things that remain hidden under an optimality account or that it is beneficial to ignore optimality and dig for a designer's – i.e., nature's – intentions instead. It is far from evident that such things can be done without entering the waters of creationism. Fortunately, the burden of proof is on, if any, those who think our understanding of biofunctions is – or should be – linked to "Mother Nature's" intentionality. (Vaesen & van Amerongen, 2008, p. 795)

Vaesen and van Amerongen's demand for a *proof* that intent reveals things that remain hidden under an optimality account is quite a stringent requirement. In the remainder of this article it should become clear that, although short of a proof, there is a significant amount of evidence that this is the case. Considerations of intention while conducting affordance-based reverse engineering of natural systems do reveal valuable things that remain hidden and unrealized under Dennett's approach.

5   Reverse Engineering and Design Recovery Techniques

If reverse engineering really is a profitable method for studying natural systems, then this approach should be fully explored to ensure that maximum benefit is achieved. A valuable resource that discusses function-based methods for hardware systems is *Product Design: Techniques in Reverse Engineering* (Otto & Wood, 2001).[16] However, a more concise, yet detailed description of the process is found in an article entitled "On Reverse Engineering" by M. G. Rekoff, Jr (Rekoff, 1985). He describes a method for systematically conducting the reverse engineering activity, noting that it "is not really greatly different from that of detective work in a criminal investigation or of conducting military intelligence operations." (Rekoff, 1985, p. 245) In a nutshell, Rekoff recommends the decomposition of existing structural hierarchy in developing functional specifications until the mechanism-of-operation is completely understood. This is unpacked in the following steps of his grand plan for conducting a reverse engineering effort . . .

- "System-engineer" [analyze the interconnectivity with an engineering mindset] first to establish hypotheses based on the information presently at hand and to identify the measurement/test needs.

---

[16]This book details function-based techniques for reverse engineering of man-made hardware. See Denis L. Feucht, "Design in Nature and the Nature of Design," (Feucht, 1999) for a discussion of the relevance of functional theories to the design of living organisms.

- Disassemble to the extent required to verify or modify the hypotheses and to perform supporting tests. One will not only uncover information relating to the hypotheses but will possibly also uncover information that was not previously known to exist.

- Further "system-engineer" on the basis of all the information in hand, form new hypotheses, and prepare for additional measurement and testing.

- Further disassemble, measure, and test to validate hypotheses and uncover new information.

This process continues until the degree of understanding is adequate for the purposes of the reverse engineering effort. For any particular item within the overall system being analyzed, "the generic process consists of the following five sequenced steps: assimilate existing data; identify elements; disassemble; analyze, test, and dimension; and complete documentation." Each of these steps is explained in very helpful detail in Rekoff's previously-referenced article.

Another helpful article, although dealing mainly with the reverse engineering of computer software, introduces the concept of "design recovery." "Reverse Engineering and Design Recovery: A Taxonomy" by Elliot Chikofsky and James Cross defines design recovery as "a subset of reverse engineering in which domain knowledge, external information, and deduction or fuzzy reasoning are added to the observations of the subject system to identify meaningful higher level abstractions beyond those obtained directly by examining the system itself."(Chikofsky & Cross II, 1990, p. 15) In other words, the goal of design recovery is to work out, at a higher level of understanding, what a system or component was engineered to do, and (to some degree of confidence) why, rather than just examining its subcomponents and their interrelationships. This generally involves extracting design artifacts by detecting design patterns, for example, and synthesizing abstractions that are less dependent on implementation. The authors believe that it is these higher level abstractions that are the key to fully reverse engineering complex natural systems. In another article, Ted Biggerstaff further expounds on the idea, "Design recovery recreates design abstractions from a combination of code [system], existing design documentation (if available), personal experience, and general knowledge about problem and application domains. . . . Design recovery must reproduce all of the information required for a person to fully understand what a program [system] does, how it does it, why it does it, and so forth."(Biggerstaff, 1989, p. 36)

A design recovery framework for mechanical components that was recently developed by engineering researchers at the University of Windsor in Canada appears to be a promising and comprehensive approach to this problem that may be transferable to natural systems. R. Jill Urbanic and Waguih ElMaraghy contend that

> there must be a methodology for recognizing design intent. The feature shapes are not arbitrary, nor is their pattern of arrangement. . . . A

form-function link needs to be established at different levels of granularity to infer the designer's intent. . . . The proposed framework includes mapping from the application domain to the functional requirements and integrates the bottom-up reverse engineering and top-down forward engineering processes and perspectives. The objective is to be able to develop a fully described model that considers the abstract design concepts and rationale along with the specific physical details. (Urbanic & ElMaraghy, 2009a, p. 195)

The framework developed in Urbanic and ElMaraghy's paper is utilized to conduct reengineering (or value engineering – improving a system based on a reverse engineering analysis) in a subsequent paper by this group. Here they claim that "gathering this [design recovery] information allows the [reverse engineer] the means to make informed decisions as to whether the current component design is adequate, or how may it be modified to add value" (Urbanic & ElMaraghy, 2009b, p. 166). In her book *Reverse Engineering*, Kathryn Ingle emphasizes the importance of design recovery data for reengineering. She writes, "This is no small amount of information if a future attempt at value engineering is to be made. Knowledge of the hows and whys of a part's design allows the [reverse] engineer to make quantum leaps in thought when seriously contemplating value engineering" (Ingle, 1994, pp. 65–66).

As researchers increasingly apply reverse engineering to living systems,[17] a natural extension will involve the application of value engineering to human beings and the fundamental problem of the human condition. *Christian* engineers/scientists, who are aware of the ultimate solution to this problem, should be ready to present this solution in a way which people can understand, especially other engineers/scientists.[18] One approach is to demonstrate the rational (though not absolutely compelling) basis for such a life-changing decision by couching it in terms of the mature fields of reverse engineering and value engineering. This strategy is enhanced further by incorporating the concept of affordance into the reverse engineering process.

---

[17]As examples see Simon Polstra et al., "Towards Design Space Exploration for Biological Systems" (Polstra, Pronk, Pimentel, & Breit, 2008) and Jamal O. Wilson and David Rosen, "Systematic Reverse Engineering of Biological Systems" (Wilson & Rosen, 2007).

[18]For more thoughts along these lines, see Harold Heie, "Christian Engineers as Faithful Witnesses to God's Redemptive Purposes for Creation" (Heie, 1999) and Walter L. Bradley, "Is There Scientific Evidence for an Intelligent Creator of the Universe?" (Bradley, 2003) and Dominic M. Halsmer, "Why Engineers Make Good Apologists" (Halsmer, 2007), and Dominic M. Halsmer, Jon Marc Asper, and Ben Zigrang, "Enhancing Science and Engineering Programs to Equip and Inspire Missionaries to Technical Communities" (Halsmer, Asper, & Zigrang, 2011), and Dominic M. Halsmer, "Application of the 'Engineering Mindset' for Worldview Evaluation" (Halsmer, 2009), and Stephen T. Frezza, "Deus Machinator: God the Engineer" (Frezza, 2009).

6   Affordance-based Reverse Engineering

In a recent article in *Mechanical Engineering*, Jonathan Maier calls for a rethinking of design theory by emphasizing that design problems begin at the system level. He claims that "individual components can be designed only after the whole problem has been understood and defined at the system level and then decomposed into subsystems and so forth, down to individual components that can be designed using the hard engineering sciences and traditional analysis methods" (Maier, 2008, p. 34; Maier, 2011). Major engineering projects have become so complex and far-reaching that George Hazelrigg has argued that design is no longer just multidisciplinary; it is *omni-disciplinary* in that any and all disciplines may be involved in the solution to a particular design problem (Hazelrigg, 1996). Maier is convinced that the complexities of systems thinking and user interactions require design engineers to move beyond the level of simply designing for component or product function. The concept that he proposes is *affordance*, which is what a system provides to a user or to another system. An (positive) affordance is what something is "good for" or "good at" and may serve as an underlying and unifying principle in the science of design (Maier & Fadel, 2001). It is important to note that the affordances of a system depend on the physical form of the system, whereas the functions of a system do not. This is a useful feature "because it allows engineers to analyze and compare the affordances of product concepts (especially at the system level) as well as of existing products for reverse engineering" (Maier, 2008, p. 36).

Although the term "affordance" has its roots in Gestalt psychology (Koffka, 1935), perceptual psychologist James Gibson defined it in the modern era as "what it [the environment] *offers* the animal [user], what it *provides* or *furnishes*, either for good or ill. . . . I mean by it something that refers to both the environment and the animal [user] in a way that no existing term does. It implies the complementarity of the animal [user] and the environment" (Gibson, 1979, p. 127). Donald Norman, in his book *The Design of Everyday Things* (Norman, 1988), steers the concept toward engineering design, but stops short of Maier's contention that it is fundamental to the design or reverse engineering of any artifact. However, Norman does admit that "affordances provide strong clues to the operations of things" (Norman, 1988, p. 9). Maier emphasizes that the affordance-based approach to design captures important interactions within the engineer-artifact-user triad that the function-based approach misses. In reverse engineering applications, a fourth role must be added to the triad—that of the investigator or reverse engineer. Interesting interactions may be elucidated among these four roles by incorporating the concept of affordances. Maier compares the strengths and weaknesses of function-based and affordance-based design and comes to an interesting conclusion; there are several "appealing synergies" or "ways in which strengths of function complement weaknesses of affordance, and ways in which the strengths of affordance complement the weaknesses of function" (Maier & Fadel, 2002). These synergies include a better view of "the big picture," a uniting of design methods, and a theoretical underpinning of design that illuminates the proper role of function.

In further work by Jonathan Maier and Georges Fadel on affordance-based methods for design and reverse engineering, they describe two kinds of affordances. Artifact-user-affordances (AUA) are defined as "the set of interactions between artifact and user in which properties of the artifact are or may be perceived by the user as potential uses. The artifact is then said to *afford* those uses to the user. Artifact-artifact-affordances (AAA) can be defined as the set of interactions between two artifacts in which some properties of one artifact interact in some useful way with properties of the other artifact" (Maier & Fadel, 2003). In addition, they explain how both types of individual affordances can be considered as positive or negative depending on whether the interaction is good for the system or not. Furthermore, each affordance can have a quality of interaction associated with it depending on how well the interaction serves its role. For example, a chair may afford sitting much better than a large rock, and the weight of a large rock may be seen as a negative affordance if portability is an issue. Maier and Fadel claim that a systematic evaluation of the affordances of each part of a system can be combined into a comprehensive affordance structure matrix (ASM), including both AUA and AAA (Maier, Ezhilan, and Fadel, 2003; Maier, Sandel, & Fadel, 2008). The system is then effectively reverse engineered in the sense that its operation is well understood. The ASM lists system components on a horizontal axis and affordances along a vertical axis. The matrix is populated by identifying which components are associated with each affordance.

7 How Affordances Assist in the Reverse Engineering of Natural Systems

In March of 2002, the cover story for *Science* was entitled "Reverse Engineering of Biological Complexity" (Csete & Doyle, 2002), which encouraged the application of engineering concepts like those borrowed from control theory to better understand the complex workings of biological systems. The article also emphasized the modularity and protocols that were being discovered within such systems at the time. Protocols were described as "rules that prescribe allowed interfaces between modules, permitting system functions that could not be achieved by isolated modules" (Csete & Doyle, 2002, p. 1666). Thus, understanding interacting modules and the protocols that specify those interactions is critical to identifying the affordances between modules (AAA). An interesting analogy was drawn between biological structures and the modules and protocols of the Lego toy system, demonstrating in multiple ways how biological systems exhibit features that are readily associated with engineered devices. The article concludes by encouraging biologists and engineers to "compare notes." Since this paper was published, it seems that many have been doing just that.

Physiologist Robert Eisenberg comments that it is most helpful to "look at biological systems through an engineer's eyes" (Eisenberg, 2007, p. 376). Emphasizing how this leads to successful experimentation, he writes, "But it seems clear, at least to a physiologist, that productive research is catalyzed by assuming that most biological systems are devices. Thinking today of your biological preparation as a device tells you what experiments to do tomorrow." Many of these types of experiments serve to identify affordances. At Stanford University, an aerospace engineer, Claire Tomlin, and a molecular biologist, Jeffrey Axelrod, teamed up to extol recent efforts aimed at "understanding biology by reverse engineering the control" (Tomlin & Axelrod, 2005, pp. 4219–4220). They summarize an extensive reverse engineering study of the mechanism used by bacterium E. coli to combat heat shock (El-Samad, Kurata, Doyle, Gross, & Khammash, 2005), claiming that it "is just what a well-trained control engineer would design, given the signals and the functions available" (Tomlin & Axelrod, 2005, 4219). Viewing the heat shock response as a control engineer would, the researchers developed a reduced order mathematical model to analyze the dynamics of each of the interacting modules. This analysis then motivated a series of simulation experiments on a larger mathematical model. The experiments involved disconnecting one or more information pathways (both feedback and feedforward protocols) which allowed comparison of various closed-loop and open-loop responses. As a result, they obtained a clearer picture of how certain information pathways and protocols afforded robustness, efficiency, and noise attenuation. Tomlin and Axelrod claim that this analysis

> is important not just because it captures the behavior of the system, but because it decomposes the mechanism into intuitively comprehensible parts. If the heat shock mechanism can be described and understood in terms of engineering control principles, it will surely be informative to apply these principles to a broad array of cellular regulatory mechanisms and thereby reveal the control architecture under which they operate . . . Certainly these kinds of analyses promise to raise the bar for understanding biological processes. (Tomlin & Axelrod, 2005, p. 4220)

8   Affordances for Human Life

In order to illustrate the application of this approach to even more complex natural systems, consider the further example of a human being. Is it possible, or even reasonable, to apply reverse engineering and design recovery techniques in an effort to gain insight into this system?[19] Of course, this approach assumes the possibility of a

---

[19]A very interesting and extensive attempt at reverse engineering human life is documented in Bryant M. Shiller, *Origin of Life: The Fifth Option* (Shiller, 2004), which finishes with the surprising conclusion that an alien intelligence is using humanity as a self-perpetuating and robust information storage system.

designer whose products lend themselves to such techniques. Multiple popular world-views adhere to the existence of a master engineer/architect of some kind, and the point was made earlier that complex natural systems are currently being successfully reverse engineered. Thus, this seems like an interesting and fruitful pursuit. This approach is especially fitting under the further assumption that humans are "made in the image" of such a master engineer, which is characteristic of a Christian world-view. This would presumably help to explain why there is such a profitable match between the complexity of the cosmos and the mental capabilities of humans. Such a match forms the basis for any successful reverse engineering of natural systems. However, the Christian worldview also entails the existence of an adversary who opposes the work of the master engineer, which may make the reverse engineering task more challenging.

The first step is to define the roles of artifact, user, investigator, and engineer that make up the big picture of design and reverse engineering. For this case the artifact can be considered as the human being, although it is difficult to separate humans from their finely tuned cosmic environment. The end user is the human mind, or consciousness, in the sense that human minds make use of their bodies in living their lives. In a Christian worldview, there is also a sense in which God is the end user, since he has made all things for himself. The investigator, or reverse engineer, is anyone who is curious enough to pursue this kind of study. It is important to recognize this so that the investigator can limit the bias that affects his or her analysis. Admittedly, there is some overlap among these roles, but that should not negate the potential fruitfulness of this analysis. As for the engineer, whether it is the tinkerer of naturalism, or God, or space aliens, it is hoped that more can be discovered about this engineer's motives, methods and intentions (or lack thereof) by properly applying the reverse engineering process.

Just listing all the affordances that the universe (artifact) exhibits towards the human consciousness (user) would be a complex and time-consuming task, but by looking at those that are the most significant, insights can begin to be gleaned. To start off, the cosmos affords humans with the ability to learn. By studying the universe and its interactions, such as those between stellar bodies, molecules, or light, humans are able to gain knowledge about why previous natural events occurred the way they did and predict how future events might follow the same pattern. The human mind is able to experience these because of the sensory organs, which the human body, as an artifact in the universe, affords to the human mind.

This brings up an interesting question; what else does the human body afford the consciousness? According to Oxford philosopher Richard Swinburne (Swinburne, 2004, p. 169), in order to be able to interact and learn about the environment in which it is placed, and yet still be a humanly free intelligence, the mind needs the human body to contain:

1. Sense organs with enormous capacity to receive information regarding distant world states.

2. An information processor to turn sense organ states into brain states (resulting in beliefs).

3. A memory bank to file states correlated with past experiences (needed to reason).

4. Brain states that give rise to desires, both good and evil.

5. Brain states caused by many different purposes that humans have.

6. A processor to turn these states into limb and other movements (needed to act on purposes).

7. Brain states that are not fully determined by other physical states (needed for free choice).

Another affordance that the universe provides the human consciousness is an environment to thrive in.[20] One key aspect of the universe is the fact that it is so finely tuned.[21] As Swinburne states,

> Not all initial conditions (ICs) or laws of nature would lead to, or even permit, the existence of human bodies at some place or other, and at some time or other, in the universe. So we may say that the universe is 'tuned' for the evolution of human bodies if the laws and initial conditions allow this to occur. If only a very narrow range of laws and ICs allow such evolution, then we may say that the universe is 'fine-tuned' for this evolution. Recent scientific work has shown that the universe is fine-tuned. (Swinburne, 2004, p. 172)

Now, the existence of the human body, due to these fine-tuned conditions is an important aspect of human consciousness because of the affordances mentioned above. What is even more amazing is that the fine-tuning of the universe not only allows for life but allows for persons who are able to question why the universe is the way it is. The fine-tuning of the universe then also affords the human consciousness with the ability to discover attributes of the engineer by questioning and observing how something so encompassing as the cosmos can be so elegant and well-engineered. With this in mind, the investigator can then appreciate the overlap with the area of the user as the human consciousness.

---

[20]In James V. Schall, *The Order of Things*, the author writes, "To be allowed 'a place on earth' wherein to work out the distinction of good and evil is what, in the end, the cosmos and earth seem to be about" (Schall, 2007, p. 83).

[21]In Alister E. McGrath, *A Fine-Tuned Universe: The Quest for God in Science and Theology*, the author makes the point that not only is the universe fine-tuned for life, but human beings are fine-tuned to profitably interact with the universe. Regardless of how this came about, the result has significant philosophical and theological implications (McGrath, 2009).

Considering that the user and investigator are roles filled by the same person brings a new dynamic and even more complex affordances between the different components of the system. As the investigator, the human consciousness begins to question why things are the way they are within the system, and if there is any way to improve the system. Humans have been attempting this from the beginning of conscious thought with innovations and inventions meant to increase the quality of life. Recently, what scientists have discovered is that many of these new inventions have already existed in nature. A prime example of this is the boat propeller and the bacterial flagellum.

The ability of humans to not only perceive affordances but also imagine how they might be creatively achieved is discussed in an article entitled "Affordances are Signs"[22] by psychologist John Pickering (Pickering, 2007). He asserts,

> The characteristic reflexivity of human cognition means that we are not only able to perceive the world as it is, that is, to perceive the affordances that actually surround us, but also to perceive affordances that do not yet exist, that is, to perceive the world as if it were otherwise. When we take a rock and modify it with blows until it functions as a blade, we do just that. We not only perceive what is, but also what may be and hence we may take meaningful, intentional action to bring it about if we so choose. (Pickering, 2007, p. 73)

The interesting thing that Swinburne points out is that the choices humans make not only tend to affect the world and others around them but also to affect their own characters (recursion), giving them some measure of influence over the kinds of people they eventually become. This is a very interesting affordance, which when coupled with the human ability to envision a preferable future, leads people to form various purposes related to self-organization (autopoiesis) and self-improvement. However, a common experience among humans is the inability to realize accepted standards and expectations. In an article entitled "The Kingdom of God and the Epistemology of Systems Theory: The Spirituality of Cybernetics" (Buker, 2009), William Buker, a Professor of Counseling, suggests that features of the human condition such as recursion and autopoiesis, combined with the nature of the universe, serve to invite humans into a more complementary coupling with the larger mind of the master engineer. He also points out how this idea of fitting with the environment

---

[22]McGrath also picks up on this theme in *A Fine-Tuned Universe: The Quest for God in Science and Theology*, writing, "The emergence of the discipline of semiotics has encouraged us to see natural objects and entities as signs, pointing beyond themselves, representing and communicating themselves. To find the true significance of things requires the development of habits of reading and directions of gaze that enable the reflective observer of nature to discern meaning where others see happenstance and accident. Or to use an image from Polanyi, where some hear a noise, others hear a tune" (McGrath, 2009, p. 3). In David Seargent, *Planet Earth and the Design Hypothesis*, the author defines a type of complexity found in nature as "transitive complexity" if it serves as a sign that points to a "larger state of affairs beyond itself" (Seargent, 2007).

rather than fighting against it is consistent with Jesus' message of the kingdom of God.

Obviously, such possibilities require the recognition of valid epistemologies beyond those of science and history, such as are found in human experiences of joy, faith, hope, and love. Some researchers are suggesting that it is time to more fully explore the wide breadth of knowledge found in the human condition if a clearer picture of mankind's purpose and place in the universe is to be achieved. Alexander Astin, Founding Director of the Higher Education Research Institute at UCLA, argues that spirituality deserves a central place in higher education (Astin, 2004). He claims that "more than anything else . . . [this] will serve to strengthen our sense of connectedness with each other. . . . This enrichment of our sense of community will not only go a long way toward overcoming the sense of fragmentation and alienation that so many of us now feel, but will also help our students to lead more meaningful lives as engaged citizens, loving partners and parents, and caring neighbors." At the beginning point of this reverse engineering project, perhaps it is too early to speculate about indications of the ultimate purpose for mankind, but it seems that Dr. Astin has put his finger on the pulse of what may be humanity's greatest desire, function and purpose—that of love. A scientific perspective of love is currently being pursued by the Institute for Research on Unlimited Love at Case-Western Reserve University. Thomas Jay Oord, theologian for the institute, has written extensively on this topic, including *Science of Love: The Wisdom of Well-Being* (Oord, 2004). He says that "to love is to act intentionally, in sympathetic response to others (including God), to promote overall well-being" (Oord, 2004, p. 75). Given the high value that humanity has placed on love throughout history, and its power to change lives for the better, it is hard to imagine that, if humans do have purpose, it would not have something to do with love. One example of a scientifically rigorous and testable model of the cosmos, which postulates the love of a master engineer and encourages humans to return that love as part of their primary purpose, is documented in the book *More than a Theory* by astronomer Hugh Ross (Ross, 2009).

9   Implications for Worldview

The advance of science in several areas is helping to identify the many affordances that the cosmos provides for higher life forms like human beings (Barrow, Morris, Freeland, & Harper Jr, 2008). These affordances are facilitated by the integration of the many subsystems of our world, which also exhibits affordances between subsystems. The connectedness of the universe is aptly described in the following quote by the famous naturalist, John Muir, "When we try to pick out anything by itself, we find it hitched to everything else" (Muir, 1911, p. 110). Some scientists have suggested that such affordances can lead to a clearer understanding of worldview, and human meaning and purpose (Denton, 1998). This is not surprising since that is ultimately what reverse engineering and design recovery procedures were designed to detect: what was the original engineer thinking? It is clear that the application of reverse engineering to natural systems cannot be separated from the worldview of the investigator because the concept of an engineer is critical to the study. Recall the importance of the relationships between engineer, artifact, and user in this regard. The authors have suggested that there is another entity that should be included in this grouping: the investigator or reverse engineer. In fact, the worldview of the reverse engineer, while playing a role in the study, may also be further shaped in the midst of the study. A classic example is the case of Charles Darwin, who, early in his career, greatly admired the works of natural theology, such as penned by William Paley, while later on, he expressed doubts as described in his writings.[23]

In teaching the process of reverse engineering and applying it to natural systems, it must be recognized that worldview plays an important role, and that worldview implications may result. The particular worldview of an investigator may or may not facilitate discovery and reverse engineering in natural systems. The extent to which a worldview is helpful in this regard is simply the extent to which a worldview is an accurate understanding of reality. Care should be taken to allow students and investigators to shape their worldviews in accordance with the evidence. Worldviews are determined based on various kinds of evidence and experiences, but scientific evidence typically plays a major role. The fact that the natural world is so readily and profitably reverse engineered suggests that the cosmos actually is an engineered system; it is the work of a Mind with extraordinary engineering expertise. Investigators should not hesitate to consider this perspective, since it not only seems to facilitate discovery but may also provide a sublimely satisfying understanding of personal meaning and purpose.

A research study involving undergraduate honors students and faculty from a diversity of academic fields, including engineering, theology, psychology, engineering

---

[23] In C. R. Darwin, *The Life and Letters of Charles Darwin*, edited by his son, Charles Darwin writes,

> I cannot see as plainly as others do, and as I should wish to do, evidence of design and beneficence on all sides of us. There seems to me too much misery in the world. I cannot persuade myself that a beneficent and omnipotent God would have designedly created the Ichneumonidae with the express intention of feeding within the living bodies of caterpillars, or that a cat should play with mice. . . . On the other hand, I cannot anyhow be contented to view this wonderful universe, and especially the nature of man, and to conclude that everything is the result of brute force. I am inclined to look at everything as resulting from designed laws, with the details, whether good or bad, left to the working out of what we may call chance. Not that this notion at all satisfies me. I feel most deeply that the whole subject is too profound for the human intellect. A dog might as well speculate on the mind of [Isaac] Newton. Let each man hope and believe what he can. (Darwin, 1887, p. 312)

physics, music, biology, chemistry, and biomedical engineering is currently underway at Oral Roberts University (Halsmer & Beck, 2012). The students engage in weekly extra-curricular research and discussions centered on the hypothesis of "God as Engineer." This hypothesis is continuously evaluated in light of a serious engagement with mainstream scientific understandings as well as a respect for traditional religious truth claims. Students assess research in the sciences, mathematics, humanities, theology, and reverse engineering methodologies for applicability to the hypothesis in question. Weekly meetings are held to discuss findings and develop journal articles, conference papers, and presentations. The diversity of academic fields represented provides rich and fertile ground for discussion and conflict resolution. Survey data from current and previous group participants suggest that these studies help students to solidify personal integrity, knowledge of purpose, and spiritual vitality, and enhance their ability to effectively communicate such convictions.

10   Conclusions

As the concept of reverse engineering is increasingly applied to natural systems, it will be helpful for students and investigators to understand the procedures for such studies. Such procedures appear to be largely absent from the current literature. A reverse engineering procedure is outlined which makes use of the concept of affordance, a recent innovation from the field of design engineering. It is suggested that affordance-based reverse engineering may be particularly useful when applied to complex natural systems. The example of human life is briefly discussed. It is recognized that investigator worldview is inseparable from such reverse engineering studies, for they are not conducted in a vacuum. Furthermore, worldview may be significantly shaped in the midst of such endeavors, since reverse engineering and design recovery activities ultimately attempt to identify original meaning and purpose. Anecdotal and survey data from undergraduate students engaged in such studies generally indicate an increased appreciation for the affordances of life and the ingenuity displayed in the cosmos. They also demonstrate an increased ability to articulate a satisfying and coherent understanding of the meaning and purpose of human life, as well as the reasons for the hope that they possess.

References

Astin, A. W. (2004). Why spirituality deserves a central place in higher education. *Liberal Education*, 90(2), 34–41.

Auffray, C. & Noble, D. (2009). Origins of systems biology in William Harvey's masterpiece on the movement of the heart and the blood in animals. *International Journal of Molecular Science*, 10(4), 1658–1669.

Barrow, J. D., Morris, S. C., Freeland, S. J., & Harper Jr, C. L., Eds. (2008). *Fitness of the cosmos for life: Biochemistry and fine-tuning*. Cambridge: Cambridge University Press.

Biggerstaff, T. J. (1989). Design recovery for maintenance and reuse. *Computer*, 22(7), 36–49.

Bradley, W. L. (2003). Is there scientific evidence for an intelligent creator of the universe? In S. Luley, P. Copan, & S. W. Wallace (Eds.), *Science: Christian perspectives for the new millenium* (pp. 159–204). Christian Leadership Ministries.

Buker, W. (2009). The kingdom of God and the epistemology of systems theory: The spirituality of cybernetics. *Zygon: Journal of Religion and Science*. Manuscript pending publication.

Chikofsky, E. J. & Cross II, J. H. (1990). Reverse engineering and design recovery: A taxonomy. *IEEE Software*, 7(1), 13–17.

Csete, M. & Doyle, J. (2002). Reverse engineering of biological complexity. *Science*, 295, 1664–1669.

Darwin, C. (1887). *The life and letters of Charles Darwin*, volume 2. John Murray.

Dennett, D. (1996). *Darwin's dangerous idea: Evolution and the meaning of life*. New York: Simon and Schuster.

Dennett, D. C. (1990). The interpretation of texts, people and other artifacts. *Philosophy and Phenomenological Research*, 50(2), 177–194.

Denton, M. (1998). *Nature's destiny: How the laws of biology reveal purpose in the universe*. New York: Free Press.

Eilam, E. (2005). *Reversing: Secrets of reverse engineering*. Hoboken, NJ: Wiley.

Einstein, A. (1987). *Letters to Solovine*. New York: Philosophical Library.

Eisenberg, R. S. (2007). Look at biological systems through an engineer's eyes. *Nature*, 447, 376.

El-Samad, H., Kurata, H., Doyle, J. C., Gross, C. A., & Khammash, M. (2005). Surviving heat shock: Control strategies for robustness and performance. *Proceedings of the National Academy of Sciences*, 102, 2736–2741.

Feser, E. (2008). *The last superstition: A refutation of the new atheism*. South Bend, IN: St. Augustine's Press.

Feucht, D. L. (1999). Design in nature and the nature of design. *Origins and Design*, 19(2). Available from http://www.arn.org/docs/odesign/od192/designinnature192.htm

Frezza, S. T. (2009). Deus machinator: God the engineer. *Proceedings of the 2009 Christian Engineering Education Conference*, (p. 30).

Fuller, S. (2006). Intelligent design theory: A site for contemporary sociology of knowledge. *Canadian Journal of Sociology*, 31(3), 277–289.

Geivett, R. D. (2005). David Hume and the cumulative case argument. In J. Sennett & D. Groothuis (Eds.), *In defense of natural theology: A post-humean assessment* (pp. 297–329). Downers Grove, IL: Intervarsity.

Gibson, J. J. (1979). The theory of affordances. *The Ecological Approach to Visual Perception*.

Graham, M. B. (1967). *Be nice to spiders*. New York: HarperCollins.

Halsmer, D. (2007). Why engineers make good apologists. In M. A. Trent, T. Grizzle, M. Sehorn, A. Lang, & E. Rogers (Eds.), *Religion, culture, curriculum, and diversity in 21st century America* (pp. 85–94). Lanham, MD: University Press of America.

Halsmer, D. (2009). Application of the 'engineering mindset' for worldview evaluation. In W. Adrian, M. Roberts, & R. Wenyika (Eds.), *Engaging our world: Christian worldview from the ivory tower to global impact* (pp. 137–152). Tulsa, OK: W & S Academic Press.

Halsmer, D., Asper, J., Roman, N., & Todd, T. (2009a). The coherence of an engineered world. *International Journal of Design & Nature and Ecodynamics*, 4(1), 47–65.

Halsmer, D., Asper, J. M., & Zigrang, B. (2011). Enhancing science and engineering programs to equip and inspire missionaries to technical communities. *Journal of the Scholarship of Teaching and Learning for Christians in Higher Education*, 5(1), 15–33.

Halsmer, D. & Beck, J. (2012). Encouraging spiritual vitality through multidisciplinary discussions on the role of engineering in reconciling science and faith issues. *Spirituality and Honors Education Symposium, Indiana Wesleyan University, May 29–31.*

Halsmer, D. & Fitzgerald, J. (2011). Metaphysical considerations enhance reverse engineering studies. Presented at the ASA Annual Meeting, North Central College, July 29–August 1.

Halsmer, D., Halsmer, N., Johnson, R., & Wanjiku, J. (2008). The applicability of engineering design principles in formulating a coherent cosmology and worldview. *Proceedings of the 2008 ASEE Annual Conference.*

Halsmer, D. & McDonough, S. (2011). Affordance-based reverse engineering of natural systems with possible corruption. *Proceedings of the 2011 Christian Engineering Education Conference (CEEC).*

Halsmer, D., Roman, N., & Todd, T. (2009b). Integrating the concept of affordance into function-based reverse engineering with application to complex systems. *Proceedings of the 2009 ASEE Annual Conference.*

Hazelrigg, G. A. (1996). *Systems engineering: An approach to information-based design.* Upper Saddle River, NJ: Prentice Hall.

Heie, H. (1999). Christian engineers as faithful witnesses to God's redemptive purposes for creation. *Proceedings of the 1999 Christian Engineering Education Conference.* Available from http://www.calvin.edu/academic/engineering/ces/ceec/1999/heie.htm

Ingle, K. (1994). *Reverse engineering.* New York: McGraw Hill.

Khammash, M., Tomlin, C. J., & Vidyasagar, M. (2008). Guest editorial—special issue on systems biology. *IEEE Transactions on Automatic Controls and IEEE Transactions on Circuits and Systems: Special Issue on Systems Biology,* (pp. 4–7).

Koffka, K. (1935). *Principles of gestalt psychology.* New York: Harcourt Brace.

Lander, A. D. (2004). A calculus of purpose. *PLoS Biology,* 2(6), e164. Available from http://www.plosbiology.org/article/info:doi/10.1371/journal.pbio.0020164

Lander, A. D. (2007). Morpheus unbound: Reimagining the morphogen gradient. *Cell,* 128, 254.

Maier, J. R. (2008). Rethinking design theory. *Mechanical Engineering,* 130(9), 34–37.

Maier, J. R. (2011). *Affordance based design: Theoretical foundations and practical applications.* Saarbrucken, Germany: VDM Verlag.

Maier, J. R., Ezhilan, T., & Fadel, G. (2003). The affordance structure matrix—a concept exploration and attention directing tool for affordance based design. *Proceedings of the ASME IDETC/CIE Conference.*

Maier, J. R. & Fadel, G. (2001). Affordance: The fundamental concept in engineering design. *Proceedings of the ASME Design Theory and Methodology Conference.*

Maier, J. R. & Fadel, G. (2002). Comparing function and affordance as bases for design. *Proceedings of the ASME Design Theory and Methodology Conference.*

Maier, J. R. & Fadel, G. (2003). Affordance-based methods for design. *Proceedings of the ASME Design Theory and Methodology Conference.*

Maier, J. R., Sandel, J., & Fadel, G. (2008). Extending the affordance structure matrix—mapping design structure and requirements to behavior. *Proceedings of the 10th International Design Structure Matrix Conference.* Available from http://the-design-works.com/pubs/abd/DSM_conf.pdf

Marchant, J. (2009). *Decoding the heavens: A 2000-year-old computer—and the century-long search to discover its secrets.* Cambridge, MA: Da Capo Press.

McGrath, A. E. (2006). *The order of things: Explorations in scientific theology.* Malden, MA: Blackwell Publishing.

McGrath, A. E. (2008). *The open secret: A new vision for natural theology.* Malden, MA: Blackwell Publishing.

McGrath, A. E. (2009). *A fine-tuned universe: The quest for God in science and theology.* Louisville, KY: Westminster John Knox Press.

Menuge, A. J. L. (2003). Interpreting the book of nature. *Perspectives on Science and Christian Faith,* 55(2), 88–98.

Meyer, S. C. (2009). *Signature in the cell: DNA and the evidence for intelligent design.* New York: HarperCollins.

Muir, J. (1911). *My first summer in the Sierra.* Boston: Houghton Mifflin, 1988 edition.

Nash, R. H. (1988). *Faith and reason: Searching for a rational faith.* Grand Rapids: Zondervan.

Naugle, D. K. (2002). *Worldview: The history of a concept.* Grand Rapids: Eerdmans.

Norman, D. A. (1988). *The design of everyday things.* New York: Basic Books.

Oord, T. J. (2004). *Science of love: The wisdom of well-being.* West Conshohocken, PA: Templeton Foundation Press.

Otto, K. N. & Wood, K. L. (2001). *Product design: Techniques in reverse engineering and new product development.* Upper Saddle River, NJ: Prentice Hall.

Pickering, J. (2007). Affordances are signs. *TripleC,* 5(2), 64–74. Available from http://www.triple-c.at/index.php/tripleC/article/viewFile/59/61

Polstra, S., Pronk, T. E., Pimentel, A. D., & Breit, T. M. (2008). Towards design space exploration for biological systems. *CIRP Journal of Computers,* 3(2), 1–9.

Ratzsch, D. (2004). Design: What scientific difference could it make? *Perspectives on Science and the Christian Faith,* 56(1), 14–25.

Reeves, G. T. & Fraser, S. E. (2009). Biological systems from an engineer's point of view. *PLoS Biology,* 7(1), e1000021. Available from http://www.plosbiology.org/article/info:doi/10.1371/journal.pbio.1000021

Rekoff, M. G. (1985). On reverse engineering. *IEEE Transactions on Systems, Man, and Cybernetics,* SMC-15(2), 244–252.

Richardson, R. C. (2007). *Evolutionary psychology as maladapted psychology.* Cambridge, MA: MIT Press.

Roberts, M. B. (1999). Design up to scratch? A comparison of design in Buckland (1832) and Behe. *Perspectives on Science and Christian Faith,* 51(4), 244–252.

Ross, H. (2008). *Why the universe is the way it is.* Grand Rapids: Baker.

Ross, H. (2009). *More than a theory.* Grand Rapids: Baker.

Samples, K. R. (2007). *A world of difference: Putting Christian truth-claims to the worldview test.* Grand Rapids: Baker.

Schall, J. V. (2007). *The order of things.* San Francisco: Ignatius Press.

Seargent, D. (2007). *Planet earth and the design hypothesis.* Lanham, MD: Hamilton Books.

Sedley, D. (2009). *Creationism and its critics in antiquity.* Berkeley, CA: University of California.

Shiller, B. M. (2004). *Origin of life: The fifth option.* Victoria, BC: Trafford.

Swinburne, R. (2004). *The existence of God.* Oxford: Oxford University Press.

Thorson, W. R. (2003). Hermeneutics for reading the book of nature: A response to Angus Menuge. *Perspectives on Science and Christian Faith*, 55(2), 99–101. Available from http://www.asa3.org/ASA/PSCF/2003/PSCF6-03Thorson.pdf

Tomlin, C. J. & Axelrod, J. D. (2005). Understanding biology by reverse engineering the control. *Proceedings of the National Academy of Sciences*, 102, 4219–4220. Available from http://www.eecs.berkeley.edu/~tomlin/papers/journals/ta05_pnas.pdf

Urbanic, R. J. & ElMaraghy, W. (2009a). A design recovery framework for mechanical components. *Journal of Engineering Design*, 20(2), 195–215.

Urbanic, R. J. & ElMaraghy, W. (2009b). Using axiomatic design with the design recovery framework to provide a platform for subsequent design modifications. *CIRP Journal of Manufacturing Science and Technology*, 1, 165–171.

Vaesen, K. & van Amerongen, M. (2008). Optimality vs. intent: Limitations of Dennett's artifact hermeneutics. *Philosophical Psychology*, 21(6), 779–797.

Wilson, E. O. (1998). *Consilience: The unity of knowledge.* New York: Knopf Publishing.

Wilson, J. O. & Rosen, D. (2007). Systematic reverse engineering of biological systems. *Proceedings of the ASME IDETC/CIE Conference*.

Wu, C. (2008). Some disassembly required. *ASEE Prism*, 18(2), 57–59. Available from http://www.prism-magazine.org/oct08/tt_01.cfm

3

The Independence and Proper Roles of Engineering and Metaphysics in Support of an Integrated Understanding of God's Creation

ALEXANDER R. SICH

Franciscan University of Steubenville

Abstract

In the *speculative* (or "theoretical") sciences—including mathematics, natural sciences, and metaphysics—the world is studied independent of human volition, calling people to recognize the truths obtained about the world as valuable in their own right. Indeed, these disciplines are ordered to "understanding-thinking" as an end in itself. The engineering disciplines, in contrast, are *productive* sciences ordered to "understanding-making"—not as ends in themselves but to achieve practical ends *per our wills*.

Natural science and engineering focus on different subject areas. In physics all forms of natural *non-living* matter in physical motion are studied by modeling "objects" according to mathematical formalisms while employing *univocal* terms such as force, energy, mass, charge, etc. In biology all natural *living things* are studied. In engineering disciplines knowledge gained from the natural sciences is applied to achieve practical ends—to the *making* of artificial *things* (artifacts). (Principles of motion of natural things are *immanent to them*, whereas artifacts' principles of motion are imposed *externally*.)

The knowledge obtained by the particular (or individual) natural sciences and engineering disciplines is limited because they *all* presuppose certain extra-scientific concepts and principles. These concepts and principles cannot be derived from any of the natural sciences themselves, for that would be circular. Moreover, the scientific method cannot validate its own ability to guide scientists to truths about creation: it cannot be the epistemic arbiter of all knowledge—otherwise known as the non-scientific pseudo-philosophy of scientism.

It falls to metaphysics to include the study of the most general principles common to all contingent beings—whether natures or artifacts. For example, it is not the reduced understanding of motion studied in physics through physical efficient causality that is studied within metaphysics, but all manifestations of change *qua* change. Metaphysics does not ask, "*how* do objects change?" but "*what* is change?" In metaphysics, reality is studied in ontological terms (hence, also employing *analogous* terms), for it must understand what being, change, substance, accident, cause, potency, act, essence, etc. are in their widest throw. Moreover, metaphysics cannot be reduced to a crude synonym for "worldview": it is a rigorous speculative science that *inter alia* animates the coordinating role a realist philosophy of nature plays for the particular natural sciences and engineering.

It falls within a realist philosophy of nature to study the most common principles of the natural sciences. To provide the foundational principles which all particular sciences and engineering disciplines presuppose, there must be a way of knowing nature whose subject matter concerns the principles and causes of natural things insofar as they are natural—that is, subject to change per principles immanent to themselves. A realist philosophy of nature therefore has the same general subject matter as the natural sciences, but it applies general philosophical (rather than specific scientific) methods to study nature, and it does not suffer the operational restrictions of methodological naturalism.

Philosophy of nature must be distinguished from philosophy of science—the latter of which includes the study of *systems of reasoning about natural things*. It must not be confused with *philosophical naturalism*, nor should it be conflated with the term "natural philosophy" as used during the Enlightenment, whose antecedents reflect a slow, incremental drift from a unified understanding of nature into the fragmentary and highly-specified particular sciences observed today.

1 Introduction

> *You arranged all things in measure and number and weight.*
> Wisdom of Solomon 11:20

Metaphysics and engineering must be carefully distinguished to understand their subject matters and proper roles in support of an integrated understanding of God's Creation, and to avoid confusion that stifles such understanding, which in turn leads to an erroneous view of reality.

Today, predominate schools of philosophy are, at best, indifferent to, but more likely positively skeptical—if not openly hostile to—metaphysical thought. Metaphysics is unfortunately not understood as the classical systematic study of being as being and the properties that apply to all beings, but a grab-bag of diverse problems (e.g., free will, the existence of God, mind-body, "eviscerated" causality, etc.), whose only common bond is that they cannot be solved by the natural sciences, phenomenology, or other hyper-specialized disciplines.

Perhaps more unfortunately, metaphysics is often reduced to a crude synonym for "worldview" or "philosophy"—a cheap, pseudo-mystical account relegated to dime-bookstores. An examination of the conference itself will provide a backdrop for this study. On the conference website metaphysics is described thus:

> In short, metaphysics *is* about the ultimate nature of reality. It includes many aspects of reality that are generally skipped over in standard physics, such as choice, creativity, morality, aesthetics, etc. While many engineers implicitly use their understanding of metaphysics when developing solutions, our goal is to move that thinking into explicit terms, so that those parts of our understanding can be better explored and systematized. Science is often bound by a methodological disregard for anything other than efficient causes. However, as engineers, our job is to include the whole of reality, and provide solutions that incorporate our entire knowledge of reality. Therefore, this conference aims at starting the discussion of how the fields of metaphysics and engineering influence each other (Bartlett, 2012).

The following comments will set the stage for the discussion that follows.

1. Indeed, metaphysics is "about" the ultimate nature of reality, but in a qualified sense. Practitioners study the principles of all contingent existents (beings) and, hence, can support reflection upon "choice, creativity, morality, aesthetics" . . . and revealed knowledge. For example, in biology extra-mental objects known as *living things* are studied; within the philosophy of nature this biological knowledge of living things is reflected on to understand what *life* is; in metaphysics what *beingness* is—what it means for a thing to exist— is studied, irrespective of whether the object considered is living or not, or whether the object is an extra-mental existent or a being of reason, etc.

2. This means there is no place for the study of choice (free will) in physics because physics has an altogether different subject matter: extra-mental material objects undergoing physical changes. Free will is not an object of study in the same sense that a neutrino is. If a physicist were to suggest the only types of existents are material objects and physical phenomena, or if he were to suggest free will can be "located" behind quantum indeterminacy, or if he

were to argue that something can indeed come from nothing, that physicist would not be "doing" physics but philosophy, and bad philosophy at that.[1]

3. Moreover, metaphysical categories and "thinking" cannot be "move[d] . . . into explicit terms, so that those parts of our understanding can be better explored and systematized." The methodologies and language employed by metaphysics are not those of engineering. The latter operates almost exclusively in univocal terms; the former also employs analogous terms, primarily because "being" is itself an analogous notion. The demand that metaphysics be reduced to "explicit terms" is a surreptitious reduction of the kinds of beings studied in metaphysics down to the ontological level of those beings studied by engineering.

4. The natural sciences are *not* "often bound by a methodological disregard for all but efficient causes"; they also incorporate physically-based material causes and a "rarified" version of the formal cause as manifested through mathematics. Moreover, the natural sciences are not constricted by a non-realist or instrumentalist interpretation of their efficacy: the natural sciences can and do provide insights into the natures of the objects being studied *as actually existing*.

5. It is most manifestly not the objective or responsibility of engineering to include the whole of reality. Engineers *make artifacts* based upon the application of highly focused natural scientific knowledge through refined *techniques* upon existing matter. Engineering cannot be looked upon to distinguish between *artifacts* and *natures*, for the objects of which it is productive can *only* be artifacts: engineering presupposes—even if inchoately expressed—that *artifacts imitate natures*, and not the other way around. *Scientism* is the pseudo-philosophical notion that the modern empirical sciences (MESs) are the epistemic arbiters of all knowledge. Imputing a similar role to engineering would result in *engineergism*, which should also be avoided.

2   Definitions and Differences of Sciences[2]

Some basic foundational questions will contribute to the manner in which knowledge is obtained in that discipline:

---

[1]Examples of scientists who stray outside their fields of competence to engage in poor philosophizing on the concept of "nothing" are Stephen Hawking (2012) and Lawrence M. Krauss (2013). See also Duke University philosopher Alex Rosenberg (2013) who denies free will, denies purpose, denies human thought is "about" anything, and unabashedly promotes scientism, physicalism, and nihilism.

[2]This section summarizes salient principles presented in sections II-IV of Adler (1978, pp. 23–167).

- What is "science?"

- What is "engineering?"

- What is "metaphysics?"

A correct response to these questions—which, at the end of the day, should provide clear definitions—is part of the broader question that serves to demarcate the bounds of these disciplines responding to its own question: To what extent do each of these disciplines span the realm of human knowledge?

Man is a *reasoning being* as reflected in the definition (i.e., the logical genus and specific difference) which properly captures the essential aspect of what it means to be a *human being*, i.e., a *rational animal*—*rational* because we are "little" *logoi* created in the Image of The *Logos*. The very first sentence of Aristotle's *Metaphysics* echoes what man is by his very nature: "All men by nature desire to know" (Aristotle, *Metaphysics*, I.980a21)[3] Moreover, we as humans are commanded by Christ not to leave our brains at the door of knowing and loving God: "And thou shalt love the Lord thy God with all thy heart, and with all thy soul, and with all thy *mind*, and with all thy strength: this *is* the first commandment" (Mark 12:30 KJV also echoed in Luke 10:27). However, the *kind* of thinking man undertakes differs importantly depending on the *telos*—the end sought. As such, there is a crucial logical distinction based on the role of truth in all human activities, i.e., the distinction of seeing man as a knower for the sake of knowing, as a doer or "acting individual," and as a maker or "builder."

Stated another way, the *kind* of thinking man does (1) as a knower for the sake of knowing (i.e., productive of knowledge) differs from (2) the thinking done to act morally, socially, or politically (i.e., productive of actions), which differs from (3) the thinking done to make things (i.e., productive of artifacts). In the sphere of *knowing*, humans are concerned with *truth as truth*; in the sphere of *doing* with *truth as action* (characterized as good and evil, right and wrong, etc.); in the sphere of *making*, with *truth as beauty* (producing things that are "well-made").

Indeed, the *true*, the *good*, and the *beautiful* are among those few but extremely important metaphysical terms called "transcendentals"—terms which apply to all contingent beings to the extent they exist, and as such are not *valuative* but ontological.[4] A fly is "beautiful," but not beautiful in the way people normally

---

[3]"*All men by nature desire to know.* An indication of this is the delight we take in our senses; for even apart from their usefulness they are loved for themselves; and above all others the sense of sight. For not only with a view to action, but even when we are not going to do anything, we prefer sight to almost everything else. The reason is that this, most of all the senses, makes us know and brings to light many differences between things." (Aristotle, *Metaphysics*, I.980a21)

[4]The transcendental terms express transcendental modes of being which, according to Fr. William Wallace, are "coextensive with being; in them being manifests itself and reveals what it actually is. Just as being is never found without such properties, so these are inseparably bound up with one another in the sense that they include and interpenetrate each other. Consequently,

characterize, say, humans or roses or paintings—not in an emotional, aesthetic, quantitative, or "valuative" sense, but in the sense of an ontological hierarchy: a human exists "more"—has a greater "claim" to existence—than a fly. It is "better" to exist than to not exist at all, and it is "better" to exist at a higher ontological level than at a lower level.[5]

A rock, in turn and in this sense, is "less" beautiful than a fly, but again not an emotional, aesthetic, or "valuative" sense. God does not "exist"—He IS Existence Itself, without a hint of potency (hence, unchanging), utterly simple, and in that sense He is Beauty Itself.[6]

So how are the relevant portions of reality demarcated for study among the various scientific disciplines? What should distinguish the particular sciences is—employing logical terms of art—the *subject matter* (sometimes called "proper object") studied within each (St. Thomas Aquinas, *Commentary on the Posterior Analytics* [of Aristotle], Lectio 21, Caput 12, 77a36–77b15). In other words, the distinction is not principally rooted in what may be devised as an *a priori* classification scheme, but in the very objects studied. The question "what *kind* of an object is that?" should lead the distinction. For example, is a "neutrino" the same *kind* of thing as a "frog," or a "Venus flytrap" the same *kind* of thing as a "triangle," or is DNA the same *kind* of thing as "design"? What this implies, of course, is that the particular sciences by themselves cannot form a basis for those distinctions, for that would be circular reasoning.

(Brief digression: at this point the reader should sense just how different the particular sciences and metaphysics are from engineering. To repeat, the former seek

---

according to the measure and manner in which a thing possesses being, it partakes of unity, truth, goodness; and conversely, according to the measure and manner in which a thing shares in these properties, it possesses being. This ultimately implies that subsistent being is also subsistent unity, truth, and goodness" (Wallace, 1977, p. 85). In typical accounts there is the existent (*ens*) itself and its essence (*essentia*), then being is said to be one, good, true (*unum, bonum, verum*) and beautiful (*pulchrum*), although St. Thomas Aquinas includes two more: thing and something (*res, aliquid*) in *Disputed Questions on Truth* (q.1 a 1). The transcendentals are ontologically one; thus, they are convertible: where there is truth, there is also beauty and goodness. The important point above and beyond the notion that the transcendentals are convertible and coextensive is that they are free from the limitations of particular *kinds* of *being*: *all* contingent beings are true, good, beautiful, etc. *to the extent they exist*. These terms "transcend" not in a "vertical" sense but in the "horizontal" sense that they apply to *all* contingent beings.

[5]This understanding also stems from a correct rejection of the gross metaphysical error known as the *univocity of being*, which holds that everything that exists "possesses" the same "level" or "claim" to (or mode of) existence, e.g., that the number 2 "exists" in the same way a carbon atom exists. Clearly, the accident of quality (e.g., "red") cannot exist without the more fundamental existence of the primary substance (e.g., this particular ball) in which it inheres. Perhaps more importantly, an electron does not "enjoy" the same level of existence as the chair under whose beingness it is subsumed, and "design" does not exist in the same way bacterial flagella exist.

[6]The ultimate aim of metaphysics is knowledge of God as the human mind can acquire. Aquinas uses analogous names to give an account of the divine attributes such as wisdom, justice, mercy, being, one, true, good, etc. See *Summa Theologiae* I.13.

knowing for its own sake; the latter seeks knowing in order to make.)

To help bring out the important distinctions (and hence provide a basis for classification), the sciences will be differentiated based on the subject matter and on the outcome or result. The first question must be, "what is science?" Classically and in its widest throw, science is 'certain knowledge through causes' or 'mediate intellectual knowledge obtained through demonstration.'[7] What this means is that theology and philosophy are also sciences, although they employ different methodologies, instrumentation, and means by which to achieve their end—truth about the real world. Just as one does not employ a telescope to study bacteria, one does not employ physics to study free will or the Transfiguration.

Now that the answer to "what does science do?" has been introduced, names can be attached to the "doing": (1) the "theoretical" or "speculative" sciences are those in which truth as truth is studied; (2) in the "applied" sciences truth in doing/action and truth in making/beauty are studied; and (3) in the methodological sciences how human reasoning leads to truth are studied.

What exactly is studied within the particular sciences is perhaps best depicted in the diagram below which relates the two major distinctions noted above. In addition, the diagram will help distinguish the speculative (theoretical) sciences through their subject matter by means of the level of abstraction.

All the particular (in the sense of "individual") sciences address different aspects of being. The question then becomes "what are those *fundamentally* distinguishing aspects or characteristics?"

For example, a discipline under the theoretical sciences is the *philosophy of nature*, under which can in turn be distinguished each of the natural sciences:

- In *physics* matter in physical motion is studied by modeling material "objects" according to mathematically-formulated deterministic mechanisms.[8]

---

[7]Science (Greek *epistêmê*, Latin *scientia*) as understood by Aristotle: *in actu: Posterior Analytics*, I:2 and II:19; *in actu exercito: Physics* I:1; see also Fr. William A. Wallace, *The Modeling of Nature: Philosophy of Science and Philosophy of Nature in Synthesis* (Wallace, 1997, p. 231); Thomas Aquinas, *Commentary on the Posterior Analytics of Aristotle*, I.2.1. For Aristotle, in contrast to the modern empirical sciences, scientific knowledge is approached from the general knowledge of proximate and unreductive "wholes" from which one works toward knowledge of "parts" more "remote" to the senses. For example, we know a horse as a primary substance before we know the atoms from which the horse is constituted: there is no necessary "roadmap" that takes one from atoms "up to" the unreductive whole known as a horse, which means a horse cannot be reduced to its parts. For Aristotle, therefore, science is certain knowledge through causes and effected by demonstration. The character of *certitude* flows from the use of *proper* causes (neither remote effects nor causes) in an argument which supports such knowledge. Moreover, it *demonstrates* (in the logical sense a *sound* argument, i.e., stronger than a *valid* argument which is merely a proof) *through causes* (i.e., not incidental principles or elements) as middle terms of a demonstrative syllogism. In other words, the middle terms in such syllogisms "cause" new knowledge. So, "mediate intellectual knowledge obtained through demonstration" is an expanded form of the definition of science as "certain [demonstrated] knowledge through causes."

[8]Note for quantum mechanics one must be careful to distinguish the mathematical formalism

Divisions of Knowledge

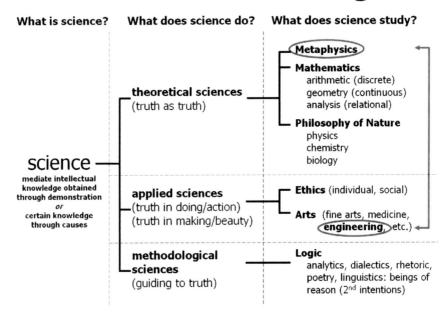

Figure 3.1: Divisions of Knowledge

- In *chemistry* the composition, structure, and properties of matter, as well as the changes matter undergoes during chemical reactions are studied.

- In *biology* living things are studied—examining their structures, functions, growth, origin, distribution, and classification.

Mathematics, as a science, stands apart from the natural sciences because it studies being from which all properties (philosophically: accidents) are abstracted except quantity—either discrete or continuous. In other words, mathematics studies being as quantified. *Metaphysics*, a science as well, stands above all—not just abstracting but *separating* everything from being except those aspects shared by *all*

---

employed to describe the behavior of objects that are highly affected by observation (measurement) from the natures of those objects. Epistemic limitations (in measurement) currently force scientists to employ statistical mathematical formalisms to describe observed quantum phenomena, but this in no way dictates the ontological status of those phenomena. The wave equation of an electron, for example, is derived from the observation of a huge population of electrons under specific conditions, but it does not follow that an electron is—by its nature—a wave: some wave-like behavior does not imply wave-like nature. An example might clarify this: the mathematical formalism known as a normal function describing the shape of the collection of balls that have fallen through a Galton board in no way implies the balls themselves—by their nature—are "spread out" in space or whose motions are *random* (i.e., without cause). The Copenhagen Interpretation commits a gross error in logic and scientific procedure by concluding that what cannot be measured exactly does not occur exactly: an epistemic principle was converted into an ontological principle.

beings. *Theology* is a science: its subject matter is knowledge of God and Divine things obtained through philosophical reflection and argumentation. *Logic* studies beings of reason or second intentions. It is the science that seeks knowledge for its own sake and is productive of something, for it organizes and guides reasoning to the truth. These examples aid the understanding of how levels of abstraction (from real beings to the object studied) distinguish the sciences—yielding the subject matter of each of these disciplines.

3 The Role of Abstraction in Distinguishing the Particular Sciences[9]

At the first level of abstraction, *particular* matter is left behind to focus upon *general* matter—where "matter" is understood in the logical sense. For example, the focus is not upon this red apple but upon the general notion of redness (physics) or the general notion of apples (biology) *universally* applied to all particular objects studied. For the philosophy of nature, any particular red apple is termed a *primary substance* while the universal apple is a *secondary substance*; redness is an *accident* (the Aristotelian category *quality*) that can inhere in many different real objects. Another example is Socrates (primary substance) vs. human being (secondary substance). This first level of abstraction—generally applied to the philosophy of nature—resolves being in the sensible and the subject matter is *ens mobile* (changeable being), while the individual natural sciences discriminate even further as noted below:

- Physics studies *natural* material "objects" undergoing physical changes

- Chemistry studies *natural* material objects subject to electron-based interactions

- Biology studies *natural* living things

"Natural"—or "natures"—as the term is used here is not to be understood in the popular or ecological sense but as applied to those things in whom the principles for motion/change are immanent. *Artifacts*, on the other hand, ultimately have their principles located externally to them. Stated philosophically, an acorn has the immanent capacity to actualize its nature into a mature oak tree.[10] A robot, which is technically termed an "accidental unity," ultimately has its motion or change located (in the sense of *explained*) as external to it: even a self-powered robot is not strictly

---

[9]Many commentaries are available on the works of Aristotle and St. Thomas Aquinas that summarize the degrees of abstraction that distinguish the speculative sciences. See, for example, Maritain (1959, pp. 35–36) and Te Velde (2006, pp. 51–53).

[10]For the distinction between natural things and artifacts: Aristotle, *Physics*, Book II, 192 b 9–18, 28. See also Stump (2006).

that, for some external rational agent had to design the robot and impart upon it the ability to be self-powered.

When physics studies motion,[11] it simplifies the things studied (termed an "object") in order to understand the motion, and then attempts to return to the fullness of reality by successively increasing the complexity of external influences. For example, a typical introductory freshman-level physics course may pose the following problem: an *object*, for which the final position is to be determined, is hurled into the air at such-and-such a speed in such-and-such a direction. It does not matter to the study of physics whether that object is a marble, a howitzer round, or an elephant. Physics does not focus on the essence ("whatness") or nature of the object, and it doesn't need to, nor can it. Once that basic level of motion is understood, one can then incorporate into the question other factors that will influence the motion: initial angular momentum, air resistance, expansion due to rapid heating, curvature of the earth, air currents, manufacturing flaws, etc. In fact, given time, patience, and resources, any level of precision can be attained. *But,* the full ontological import and essence of the "object" will never be captured by physics alone—let alone by mathematics.

The second level of abstraction belongs to mathematics: all sensible aspects are left behind except the quantitative, and hence the beings studied are resolved not in the senses but in the imagination. Even highly abstract mathematical concepts (Lie groups, Fourier analysis, complex functions, etc.) must ultimately be reducible to discrete and continuous quantities abstracted from real objects. Mathematics is therefore the study of things that can be imagined and conceived without matter, not just the abstraction from the particular for the philosophy of nature and the individual natural sciences. The focus of mathematics is upon the first Aristotelian category of real being—quantity, again whether continuous (surfaces, volumes, etc.) or discrete (integers, countables, etc.) or relational (equations, inequalities, etc.).

Quantity, as the first accident of real being, is the basis for quality—the second accident of real being. For example, the average temperature of air in a room presupposes a non-point-like *res extensa*—a thing spatially extended—and the ability to measure (quantify). Furthermore, quality admits of degree; quantity does not. For example, no number (quantity) of kindergarteners adds up to the intelligence (quality) of Einstein.[12] Finally, geometry, for example, is not concerned with the question

---

[11]To be quite precise, in physics one does not study motion in the same sense that in biology one does not study life. Neither motion nor life are, respectively, the objects of these sciences. In *physics* one studies material *things in motion*; in *biology* one studies *living things* (or *things* that were once living). "Things" here are understood as real, extra-mental existents accessible to (observable by) the five external senses or the senses enhanced through instrumentation. Neither life nor motion are *per se* observable by the senses: motion and life are both known to exist, but only because of reflection upon the knowledge gained through the natural sciences, and that reflection properly belongs to the philosophy of nature. See, in particular, section 4, Metaphysics: The Foundational Science, page 50ff ahead.

[12]It also makes for good jokes among scientists, e.g., two is equal to three for large values of two.

of whether a triangle (as imagined from some extra-mental object) is made of copper or of wood, but only with its absolute quantifiable nature, according to which it has three sides and three angles that add to 180 degrees in Euclidean (flat) space.

The final level of abstraction is actually *separation*: all material aspects are left behind or separated from that which can exist without matter. Angels, for example, are purely immaterial beings and as such cannot in any way be imagined, for imagination requires an image—a picture—in the mind. Metaphysical terms such as unity, substance, soul, potency, causes, etc. exist without matter as well because such concepts apply to all extra-mental existents. In metaphysics, therefore, one does not study being as changeable (philosophy of nature) or as quantifiable (mathematics) but being as being, and hence such being can only be resolved through concepts in the mind. As such, metaphysics is the study of the foundational principles of all beings—those aspects shared by *all* contingent beings.

In a strong sense, metaphysics is really at the heart of philosophy since it deals with the nature of reality in its widest throw. Metaphysics comprises two main areas or sub-disciplines: ontology and transcendental metaphysics. Ontology concerns itself with responding to the question "what exists?" while Transcendental Metaphysics concerns itself with the questions "what is it for something to exist?," "are there different modes of existence?," and "if there are different modes of existence, what are the truth makers for these modes?"

More specifically, *ontology* attempts to formulate a complete list of the fundamental categories of being, while *transcendental metaphysics* concerns itself with (a) a fundamental understanding of essence, existence, nature, cause, etc., and the relationships between them, (b) the truth makers for modal claims, and (c) the transcendental attributes of being—attributes which apply to all beings simply as beings.

An example of this philosophical approach is shoes. What is common or universal to all shoes cannot be drawn or imagined. As such, there is absolutely nothing which can be sensed—and hence measured—about "shoeness." When one has an idea (not image!) of what is common to all shoes, of every shape, size, color, style, etc., one has grasped *conceptually* the form (formal cause—the "whatness")—the shoeness.

4 Metaphysics: The Foundational Science

So, metaphysics is the foundational philosophical discipline—quite literally *the* foundational science. Whereas the modern empirical sciences (MESs) are the most fundamental form of knowledge because their objects are accessible through the senses, the MESs are neither the only form of knowledge nor the most important. All the particular (individual) sciences—including the productive sciences such as medicine and engineering and architecture—depend upon metaphysics for their ultimate presuppositions and basic principles. The particular sciences seek to understand their particular subject matters through the proximate "ultimate" causes within their particular domain. And, no science can prove its own proximately considered first principles. There must be one foundational science (metaphysics) which seeks to understand all reality—all contingent beings—in terms of the universal properties of being as such.

For example, *change*, and the species of change called "local motion" (translational, vibrational/oscillatory, and rotational/circular) can be illustrated through the following scenario. Suppose a group of one hundred people, chosen at random, were asked if they were seismologists that have modeled tectonic subduction zones. The expected response would be one, perhaps two. If the same group was asked if they had experienced an earthquake, the response might increase to about fifteen people. If they were asked if they know *what* motion is, or at least if they had experienced it, the hands of all the people in the group would be raised. This example highlights the difference between the narrowly, yet deeply focused work of scientists, those who have had special experiences and hence knowledge, and the common knowledge shared by all individuals.

Motion (change of position of a material object as a function of time) for a physicist is relatively easy compared to the general notion of what motion is. Physics only asks "how do things change/move?" whereas the philosopher must ask "***what is*** change in its widest throw?" Motion for the philosopher is merely a species of change, for it is not merely a metaphor to assert, "I was *moved* by the beauty of my wife." There was a "before" when I was not moved, and then there was an "after" when my potential to be moved was actualized into reality—the reality of actually being in love. Hence, the most general definition (i.e., not the narrow physics-based definition) of motion is "the reduction from potency to act, insofar as the object is in potency."[13] It is this general definition—based upon common experience accessible to all—that provides the foundation for physicists to do their good, mathematically described work. Change had to be understood in its widest, most general throw before physics could narrow that understanding of change to study material objects in physical motion.

So, the three speculative sciences can be defined as follows:

- *Physics* includes the study of those things (material objects and physical phenomena) that have separate substantial (real, i.e., extra-mental) existence but are subject to change.

- *Mathematics* includes the study of those things that are free from change but which do not have separate substantial (real, i.e., extra-mental) existence: they exist only as distinguishable aspects of concrete realities, e.g., a tire's shape is a circle, the statistical distribution of students' test scores is described by a normal curve, etc.

---

[13]According to Aristotle, motion must be defined as "the act [entelechy] of that which exists in potency insofar as it is such" (*actus existentis in potentia secundum quod huiusmodi*) (*Physics* 3.2, 201a27–29).

- *Metaphysics* includes the study of those things that have separate substantial existence but are free from change—primarily the study of substances (as they are the primary mode of being); in other words, what aspects do all substances share?

5 Quo Vadis, Engineering?

The metaphysician asks, "What is motion?" The physicists asks, "How do levers move under the influence of external forces?" The engineer asks, "How do I construct a structurally sound bridge that will not move (within set parameters) under the influence of external forces?" The differences between the questions asked and hence the objects studied by these disciplines are profound.

The goal of the speculative (theoretical or "pure") sciences is to conform one's mind to reality. The goal of the applied sciences is to conform one's actions to reality, and therefore by them know (a) which actions conform to reality, (b) what ought to be done, and (c) how and why one should act.

In fact, the study of the applied sciences can only be successful to the extent the pure sciences are incorporated. Ethics or moral philosophy answers the questions, "what should humans do" and "why?" The arts (including engineering as productive of ARTifacts) are subordinate to ethics. Engineers must first decide what ought to be done (which implies knowing why). Then and only then may they enter into the details of how to do it. For example, knowing how to abort a child cannot trump knowing whether or not one should do so. Through science and technology, weapons of mass destruction in Iran can be detected through remote sensing. But no natural scientific, technological, or engineering discipline can then determine what should be done about those weapons.

The applied sciences are subordinate to the speculative sciences as the speculative sciences are subordinate to the science that provides their principles—metaphysics. The knowledge obtained by the natural sciences must precede the application of the knowledge through engineering, and in that sense engineering is subordinate to the natural sciences as well.

Engineering can now be examined according to its relationship to the speculative sciences.

- *Metaphysics* is the study of being as being.

- The *modern empirical sciences* are the studies of the metrical properties (quantified observables) of changeable, material objects—objects particular to each natural science.

- The *philosophy of nature* stands between these two disciplines as the study of all changeable beings—of the general properties of all real (extra-mental, substantial and accidental) existents.

- The *engineering disciplines* apply the knowledge gained from the natural sciences to achieve practical ends—to the making of artificial things (artifacts— *not* natures).

- The *philosophy of science* is the study of systems of reasoning (methodological epistemology) of natural things.

So, metaphysics is ultimately connected to engineering because it includes the study of the first causes and principles of all beings—including those things studied by the modern empirical sciences and engineering.

However, the modern empirical sciences are "closer" to metaphysics because

1. They are speculative (theoretical) in the same way metaphysics is.

2. The modern empirical sciences provide real-world knowledge for reflection within metaphysics.

3. Metaphysics, through natural philosophy, provides the modern empirical sciences with foundational (proximate "first") principles.

In contrast, engineering is "located" further from metaphysics because

1. It is applied knowledge (productive of artifacts) rather than speculative.

2. It depends upon the MESs to provide knowledge of the real world.

3. It provides metaphysics no direct substantive knowledge of the real world upon which to reflect.

4. It is subject to the constraints of the practical sciences (e.g., ethics).

5. Metaphysics, through the philosophy of nature and the modern empirical sciences, provides engineering with foundational principles.

Productive reasoning involves having what one might be tempted to call "creative" ideas. In fact, productive ideas are based upon some understanding of the forms that matter can take, supplemented by imaginative thinking about such details as sizes, shapes, and configurations. Scientific knowledge can indeed be applied productively: scientific knowledge, through *technology*, provides the skills and power to produce things.

But the philosophical reflection that improves the common-sense grasp of the physical world provides neither the skill nor the power to produce anything. Philosophy bakes no cakes and builds no bridges. So, is it "useless" in the Marxist sense? No. Echoing more strongly a previous point, philosophical knowledge lays the foundations of thinking and can be "used" to direct lives and manage societies: philosophy animates for the sake of doing—not directly for the sake of making, which is why *making* must be subordinate to *doing*.

There is also an analogical connection of metaphysics to engineering insofar as metaphysicians can draw analogies from the way things are made and function to make inferences about those things that cannot change. This reflects Aristotle's crucially important principle: *Art(ifact) imitates nature*, and not the other way around.

To anticipate a possible criticism: some may point to the Ancient Egyptians as an example of those who did not know physics yet built the pyramids and statues, or Ancient Romans who also did not know physics yet built roads and aqueducts, so that engineering *can* precede science. This is a misunderstanding of the principle. In the order of knowing, the builder must first understand—even if inchoately—that rocks fall when released and crumble when too high an external load is imposed. Such knowledge must precede the application of that knowledge. To build something is not to do engineering, in the same sense that technology cannot be equated to science. To categorize the people of those times as scientists or engineers is actually incorrect if science and engineering are understood to be what they are to moderns, unless one does so analogously. This is not to suggest that engineering does not inform the natural sciences—it does quite significantly. However, there is *no reason whatsoever* for engineering to explore physical reality other than to *make* something—an artifact. If engineering's goal is simply to know, then it is no longer engineering but physics or chemistry or whatever natural science. Engineering's *telos*—its end or goal—for knowing reality is to make something, not to know simply for the sake of knowing.

6   A Case Study in the Failure to Properly Distinguish: Intelligent Design

A mistake often made is to forget (or neglect, perhaps even intentionally) this principle, and hence to fall into one of the most common errors: the fallacy that nature imitates art(ifact)—the "mechanistic" reductionist error shared by Intelligent Design and DarwinISM.[14] Consider the following claims:

---

[14]Intelligent Design proponents have attempted to deny their project is mechanistic in its view of living organisms or their constituent systems and parts, but in doing so they appear to miss the artifact/nature distinction, which is not so much a characterization of living things as machines (although some do make this error) but more deeply what is meant by mechanism is a lack of immanent powers based on living things being *per se* natures, and that therefore descent with modification must somehow be "guided" by an *external* artificer rather than having the ability to evolve on their own. See, for example, Logan Paul Gage (2010) and Jay W. Richards (2010c). Note that these gentlemen believe (as do most Intelligent Design proponents) that "design" can be empirically detected, and hence is something directly accessible to the natural sciences. It is this notion that animates the Intelligent Design movement's desire to inject its arguments into the biology classroom. If, on the other hand, Intelligent Design is correctly understood as a philosophical interpretation of scientific findings, the project must seek another venue other than biology classrooms.

1. Intelligent Design proponent Michael Behe: "Modern science reveals the *cell is a* sophisticated, automated, nanoscale *factory*" (Discovery Institute, 2007) [emphasis added].

2. Darwinist Bruce Alberts: "The entire *cell can be viewed as a factory* that contains an elaborate network of interlocking assembly lines, each of which is composed of a set of large protein machines. . . . Why do we call the large protein assemblies that underlie cell function protein *machines?* Precisely because, like machines invented by humans to deal efficiently with the macroscopic world, these protein assemblies contain highly coordinated moving parts" (Alberts, 1998, p. 291) [emphasis added].

This is muddled thinking that flips reality on its head. Behe's claim is all the worse because it is a categorical assertion, while Alberts at least qualifies his claim with "can be viewed as."

The nature or essence of a bacterial flagellum, for example, might be considered from the *artificial* (note the root "artifact") perspective as that of a motor (a machine—an accidental unity) *but only if considered in isolation from the cell.*[15] However, it is not Behe's empirical observations that indicate to him that the flagellum is a motor: he drew upon empirical observations to reason to the intelligible aspect of that thing known through the universal concept of "motor," i.e., it required an immaterial nous (intelligent agent) to "know" the immaterial intelligible aspect of that particular biological system.

Ric Machuga pointedly explains Behe and Dembski's attempts to quantify an object's form (formal cause or specificity) and *telos* (final cause or function) by providing the example of pincers. Neither mathematics nor the natural sciences can explain what the tweezers in my hand, the pliers in my tool box, the bigger pliers in my plumber's tool box, the pincers of a lobster's claw, the forceps of a surgeon, the mandibles of a bull ant, the jaw of a human being, or a fireman's "jaws of life" are. Yet, any intelligent agent can explain that all these examples are understood by the *universal (i.e., not concrete and hence immaterial) concept* of a compound lever. There is *nothing* about *what* a compound lever *is* that physics or mechanical engineering can tell anyone—apart from *how* one works: the moment arms involved, the pressure applied at the point of contact, the position of the fulcrum, etc. Natural philosophy, on the other hand, not only permits one to "see" the *quiddity* ("whatness") of the object, but in the list provided it can distinguish between the three examples of *per se* unities and the five examples of *per accidens* unities. Machuga correctly concludes: "The only thing all pincers have in common is an intended purpose, design,

---

[15]The point of the flagellum "considered *in isolation* from the cell" is also important because it highlights another form of reductionism in Behe's thinking. To study most of the aspects of a bacterial flagellum (and certainly to study the components of the flagellum and their properties), the cell must be destroyed (Nature must not only be interrogated but tortured as well–harkening back to Francis Bacon's *novus ordo seclorum* control over Nature), i.e., the flagellum is studied in a pathological state removed from the *substance* known as the bacterial cell.

or function—in short, a form. While quantifiable shapes can be the embodiment of forms, no form can be reduced to a quantifiable shape" (Machuga, 2002, p. 162).[16]

While provocative, it is nonetheless true: Intelligent Design is neither a science nor engineering. Intelligent Design is a philosophical interpretation of scientific findings with theological implications,[17] which the following example will help clarify. Physicists inferred the existence of a material object—the neutrino—from physical observables, i.e., from the apparent violation of conservation of energy and angular momentum in beta decay processes. Intelligent Design theorists attempt to infer the existence of "design" and hence a designer. But this, of course, begs the question: Is design the same ontological kind of thing as a neutrino? No, for unless one is a materialist or other such reductionist (which is a hidden irony of Intelligent Design), design is a very different kind of thing than a neutrino. This, in turn, then demands a very different *kind of science* to study design—not a natural science, but a philosophical science.

This confusion arises precisely because Intelligent Design proponents fail to distinguish properly their subject matter, i.e., they haven't asked the question: "Is a neutrino the same *kind* of thing as 'design?'" Since Intelligent Design proponents ontologically "flatland" (or "domesticate") design down to the level of material objects, they therefore attempt to measure and quantify formal and final causality. And, the theory also unnecessarily pits efficient causality against final causality.

The following may help shed some light on the situation:

1. From sensory-accessible data (effects) one can *infer* (inductively reason) to the existence of an object *as the cause of the effects*, whose existence can then be verified directly through the senses (or senses enhanced with instrumentation). The character of such an inference is *natural scientific*, and an example of such an object is the neutrino.

2. From sensory-accessible data (effects) one can *infer* (inductively reason) to the existence of an object *as the cause of the effects*, whose existence can *not* be verified directly through the senses, but which must be argued to through *philosophical inference* (philosophically if rigorously undertaken, or at least "recognized" if not). Examples include "objects" such as design and intelligence.

---

[16]Note for pedagogical reasons Machuga uses the term "shape" to capture all nine accidents (Aristotelian categories) of real being: "'Shape,' as we are using the term, refers to the totality of a thing's physically quantifiable properties, i.e., its physical shape and size, height, weight, chemical composition, etc., in its most complete description. . . . It is what the sciences study" (Machuga, 2002, p. 27).

[17]See, among many examples, the subtitle of the book and the entire first chapter of William Dembski, *Intelligent Design: The Bridge Between Science & Theology* (2002, pp. 25–48); "At the same time, contemporary ID arguments represent a return to a very traditional and biblical way of speaking" in Richards (2010c, p. 250).

3. From sensory-accessible data (effects) one can *infer* (inductively reason) to the Pure Act of Existence (*Ipsum esse subsistens*) or Ultimate Cause of all, but only through *metaphysical inference*.

What Intelligent Design attempts to do is bypass philosophical inference (to "obtain" design) and metaphysical inference (to "obtain" God). This is not possible: there is no way to argue *directly* through the natural sciences to "design," and much less so to God. The only possible means by which we can reason to the Existence of Pure Existence (God) is an argument that (a) begins with sensory-accessible effects and then argues through metaphysical reflection to higher verities (design), which culminate in the Ultimate Cause (God).

Design is an abstraction based on sensory observations, and philosophical arguments are needed to infer to its existence, just like philosophical arguments are used to support the efficacy of scientific method. A neutrino, on the other hand, is not an abstraction; it is a real, substantive, extra-mental material object. Design is a combination of formal causality (whatness) and final causality (function or "for what") that partly explains the existence of artifacts—*not* natural things.

Functionality or "irreducible complexity"[18] as understood by Michael Behe is, in reality, mathematically ambiguous, and in no way can it be measured—despite protestations to the contrary. Functionality implies a *telos*—a "for whatness" or final cause, which is neither a subject matter (formal object) nor material object studied by any of the modern empirical sciences. To repeat Machuga's point above, even if one considers a pair of pliers, a nut cracker, an insect's mandibles, and a lobster claw, as having one and the same *functionality* applied to different situations (to hold tight and possibly crush an object), their complexities vary widely. Which of these objects has more "*functionally* specified information"? Stephen Meyers explicitly claims that "Dembski's theory . . . applies to any system that has large amounts of such functional information" (Meyer, 2009, p. 372). What does "functional information" mean? Are jaws-of-life more *functional* than a pair of household pliers, or are they simply more *complex*? How exactly is one to measure—literally to quantify by means of a metric—the differences in function between the physical examples above? Again, one cannot mathematically distinguish functions. The function is the same for all the examples above (and many more), while their mathematically describable complexities (through the correlation of sensory-accessible properties into mathematical formalisms) vary widely. Complexity and information are quantifiable, while functionality and essence (meaning) are not.

Similarly, what a thing is—or, to employ William Dembski's term, its "specified complexity"[19]—is also mathematically ambiguous, and Intelligent Design actu-

---

[18]Behe defines an irreducibly complex system as one "composed of several well-matched, interacting parts that contribute to the basic function, wherein the removal of any one of the parts causes the system to effectively cease functioning" (Behe, 2006, p. 39).

[19]For a fair summary, including definitions, mathematical treatment, criticisms, and references, see Specified Complexity (2013).

ally fails to properly distinguish "information" (which is measurable) from "meaning" (which is not measurable). Intelligent Design appears to conflate the two. To be "specified" is to be a *specific* thing, to be an actualized existent—a *primary substance* that is *understood* or intelligible or whose whatness is known. In other words, it is the formal cause of a thing Dembski is attempting—and failing—to quantify. If the Ukrainian word ЧЕРВОНЕ is written with smoke in the sky in all–caps, bold–faced, with letters each 100 meters tall, and if the English word "red" is printed in small–case 8–point font size, these two words have (gender signification notwithstanding in Ukrainian) *precisely the same meaning.*[20] Yet, the information content required to represent the physical aspects of these two words in a computer's memory would be very different. Nonetheless, Dembski claims (which the scientific community correctly rejects, even as it only partly understands why): "If there is a way to *detect* design, specified complexity is it" (Dembski, 2007, p. 19) (emphasis added).[21]

Note also the apparent conflation of the terms "detect" and "infer": if design is "detectable," then direct sensory access (or senses enhanced through instrumentation) is possible; if design is "inferred," then that inference must be *philosophical* given the *kind* of thing design is, as noted earlier. If Intelligent Design is to be considered part of the natural sciences, it must be evaluated on its scientific merits, for which the track record (so far) has not been very good. If Intelligent Design is a philosophical interpretation of scientific findings, it has no more place in the biology classroom than DarwinISM does, and it must be evaluated on its philosophical merits.

Finally, the theory of Intelligent Design presupposes that living organisms are artifacts with externally imposed functionality and form. These organisms are then implicitly viewed as passive, mechanistic receptors of an occasionalist designer's tinkering.[22] As such, it fails to consider the Biblical account that God does not "make"

---

[20]Both words have differing and measurable "amounts" of information which can be understood as reflecting complexity. Yet, both words have exactly the same specification (intentionality): "redness." What metric has Dembski proposed to measure and distinguish *per se* specification? What exactly does it mean for something to be *metrically* "more" or "less" specified? The "meaning" of something is expressed through a definition (genus and specific difference) which yields the *essence*— *what* something is, which is utterly inaccessible to being measured in any way. The form is the philosophically causal explanation of *what* something is, and it *informs* our knowing *what* that object is. The *essence* is the actual *whatness* of the object expressed in a definition: it is what our minds "become" when we come to know the object. We know the meaning—what something is—when our immaterial minds are informed *essentially* to "become" that object.

[21] "Intelligence" is not an object inferred to directly *through* the natural sciences but *through* philosophical reflection. All people recognize intelligence immediately when they "see" it. Yet, that "recognition" is not "seen" optically but by the operation of the mind above and beyond the sensory-accessible import. When formalized, the "seeing" of "intelligence" is properly called "philosophical reflection."

[22]With the goal of reconciling Intelligent Design with Thomism, several authors have attempted to address these issues in Jay Richards, *God and Evolution* (2010a), chapters 9–13. Such attempts have not met with much success, in part because a number of questionable claims that are made. These claims merit a rigorous response which, unfortunately, is outside the immediate scope of this paper. For extended accounts criticizing the attempted reconciliation of Intelligent Design with

individual things in nature like some Cosmic Artificer. Rather, God *creates and maintains in existence* individual natures—natures which actualize, through their immanent capacities or powers—to perfections (understood philosophically) like the earlier example of the acorn growing into a mature oak tree: God doesn't "make" oak trees (note the inherent engineering reductionism of a natural thing to an artifact), acorns "make" oak trees. As reflected in Genesis 1:11–24 (KJV), "the *earth* brought forth" and "the *waters* brought forth". Intelligent Design demotes God's timeless, existence-sustaining creative act to some "thing" accessible to natural scientific inquiry.

At the end of the day, Intelligent Design claims that "design" and "intelligence" are "empirically detectable,"[23] and that by strong implication the existence of God can be inferred, never quite coming clean on the precise character of the inference (explained above). This is what is meant by Intelligent Design seeking direct[24] access to God's existence by means of natural scientific means: God is seen as an *external* "guider" of natural selection, which reflects an occassionalist notion of the source of actions of biological organisms. In other words, God is "domesticated" down to being a cause among causes.

As St. Paul instructs, because human beings are by nature rational animals, humans are capable of knowing "God's eternal power and divinity" from "the things He has made" (Romans 1:19–20 NASB), i.e., by the light of human reason alone:

> For what can be known about God is evident to [the wicked], because God made it *evident* to them. Ever since the creation of the world, His *invisible* attributes of eternal power and divinity have been able to be *understood and perceived* in what He has made." (emphasis added)

The knowledge presupposed by St. Paul in this passage cannot possibly be of beautiful galactic nebula, bacterial flagella, cosmic fine-tuning, or engineering system analysis of cell processes—there were no modern empirical sciences at that time! Rather, this knowledge must be grounded in pre-scientific common experience and

---

Thomism, see Edward Feser. For instance, Feser (2013, pp. 724, 745) gently chastises Jay Richards for suggesting Aquinas' views are close to those of William Paley and points to why Thomists are generally strongly opposed to Intelligent Design. In addition to his published works, Feser maintains a blog that often covers similar subjects at http://edwardfeser.blogspot.com/.

[23] "This chapter argues that God's design is also accessible to scientific inquiry. The crucial point of this chapter is that design is empirically detectable, that is, we can detect it through observation" (Dembski, 2002, p. 17)

[24] The previous footnote provides only one example of Dembski's claim that "God's design" is directly empirically detectable (that is, "we can detect it through observation") to the natural sciences, and yet in other places he denies this (Dembski, 2002, pp. 107–108). See also Richards (2010b, p. 205): "Most ID arguments are consistent with a variety of views about how and when God acts in nature. In fact, ID arguments are not explicitly theological. One could pursue research within an ID framework without ever asking follow-up questions; nevertheless, these arguments have positive theological implications. As a result, they can be used to construct arguments for God's existence."

philosophical reflection animated by the presuppositions and intellectual habits whose origin is in a Christian world view.

One would have to be out of one's mind to deny design in the universe. But, the reductionist view of "design" proffered by Intelligent Design proponents and "design" as understood by critically thinking philosophers are two different things. Reason alone and independently (as well as faith alone) can lead us to knowledge of the "existence" of Pure Existence Itself ("I AM THAT I AM," Exodus 3:14, KJV), i.e., to knowledge of the "existence" of the Creator. But, for the former the path is *through* metaphysical reflection *starting from* the natural sciences (or engineering or general observations). The former establishes a "that" for "existence" (the so-called god of the philosophers), while the latter establishes the relational "who" and is the basis for the trust known as "faith" in *the* God.

No amount of empirical data can defeat a metaphysical proof or principle. Clearly, empirical data, scientific theories, and engineering expertise *inform* metaphysical reasoning. But such reasoning is impervious to empirical falsification—similar to a mathematical theorem not being falsifiable by any empirical data. For example, no new empirical data will ever alter the fact that in the context of Euclidean (flat) geometry, the sum of all interior angles of any triangle adds to 180 degrees. Moreover, no new empirical data will ever defeat the Principle of Non-Contradiction, poor philosophizing on the part of certain quantum physicists notwithstanding.

No empirical data or scientific theories exist in a vacuum: there is no "preferred" metaphysical interpretation attached. Yet, there must be interpretations. Whether implicitly or explicitly, secularist scientists embrace the notions of "philosophical" naturalism and materialism as animated by scientism. To counter this, the Intelligent Design movement cannot hope to misappropriate science, or especially to find a "new home," in engineering. It must adopt a realist philosophy of nature. That is where the real battle for the mind is fought—not in or through the natural sciences. What this implies, of course, is that Intelligent Design proponents must face some deep soul-searching to realize Intelligent Design does not belong in the biology classroom any more than DarwinISM does. That is the only way the Intelligent Design movement will reclaim the high ground, and only then will it be unstoppable.

The failure to distinguish the proper roles and subject matters studied in the natural sciences, metaphysics, and engineering will only lead to confusion—of which Intelligent Design and DarwinISM are only two unfortunate examples.[25] However, when these disciplines are respected and their differences celebrated (a tuba would sound ugly usurping the part of a piccolo), i.e., if they are permitted to operate as a symphony productive of truth—with the particular science as the individual instruments, metaphysics as the conductor, and God as the composer—the music would, indeed, be Divinely beautiful.

---

[25]See Meyer (2009, p. 372) for an example.

References

Adler, M. J. (1978). *Aristotle for everybody.* New York: Macmillan.

Alberts, B. (1998). The cell as a collection of protein machines: Preparing the next generation of molecular biologists. *Cell,* 92, 291.

Bartlett, J. (2012). About the conference. *Engineering and Metaphysics 2012 Conference website.* Available from http://www.blythinstitute.org/site/sections/30

Behe, M. (2006). *Darwin's black box: The biochemical challenge to evolution.* New York: Free Press.

Dembski, W. A. (2002). *Intelligent Design: The bridge between science & theology.* Downers Grove, IL: InterVarsity Press.

Dembski, W. A. (2007). *No free lunch: Why specified complexity cannot be purchased without intelligence.* Lantham, MD: Rowman & Littlefield Publishers.

Discovery Institute (2007). Interview with Michael Behe on *The Edge of Evolution. Discovery Institute.* Available from http://www.discovery.org/a/4097

Feser, E. (2013). Between Aristotle and William Paley: Aquinas's fifth way. *Nova et Vetera,* 11(3). English Edition.

Gage, L. P. (2010). Can a Thomist be a Darwinist? In J. W. Richards (Ed.), *God and evolution: Protestants, Catholics, and Jews explore Darwin's challenge to faith* (pp. 187–202). Seattle: Discovery Institute Press.

Hawking, S. (2012). *The grand design.* New York: Bantam.

Krauss, L. M. (2013). *A universe from nothing: Why there is something rather than nothing.* New York: Atria Books.

Machuga, R. (2002). *In defense of the soul: What it means to be human.* Grand Rapids, MI: Baker Book House.

Maritain, J. (1959). *The degrees of knowledge.* New York: Charles Scribner's Sons. Translated by G. B. Phelan.

Meyer, S. C. (2009). *Signature in the cell: DNA and the evidence for Intelligent Design.* New York: Harper Collins.

Richards, J., Ed. (2010a). *God and evolution: Protestants, Catholics, and Jews explore Darwin's challenge to faith.* Seattle: Discovery Institute Press.

Richards, J. W. (2010b). Straining gnats, swallowing camels: Catholics, evolution and intelligent design, part i. In J. W. Richards (Ed.), *God and evolution: Protestants, Catholics, and Jews explore Darwin's challenge to faith* (pp. 203–224). Seattle: Discovery Institute Press.

Richards, J. W. (2010c). Understanding intelligent design: Catholics, evolution, and intelligent design, part iii. In J. W. Richards (Ed.), *God and evolution: Protestants, Catholics, and Jews explore Darwin's challenge to faith* (pp. 247–271). Seattle: Discovery Institute Press.

Rosenberg, A. (2013). *The atheist's guide to reality: Enjoying life without illusions.* New York: W. W. Norton & Company.

Specified Complexity (2013). In *Wikipedia.* Available from http://en.wikipedia.org/wiki/Specified_complexity

Stump, E. (2006). Substance and artifact in Aquinas's metaphysics. In T. Crisp, M. Davidson, & D. Laan (Eds.), *Knowledge and reality: Essays in honor of Alvin Plantinga* (pp. 63–79). Dordrecht: Springer Netherlands.

Te Velde, R. A. (2006). *Aquinas on God: The "divine science" of the* Summa Theologiae. London: Ashgate Publishing, Ltd.

Wallace, W. A. (1977). *The elements of philosophy.* New York: Alba House.

Wallace, W. A. (1997). *The modeling of nature: Philosophy of science and philosophy of nature in synthesis.* Washington, DC: Catholic University of America Press.

Part II

Architecture and the Ultimate

Architecture is one of the few engineering disciplines which has had a deep and explicit connection with philosophy and theology, largely because of the religious significance of many important structures. Modern architecture has lost many of its roots, tending almost entirely towards a functionalist view of itself. Mark Hall, using Ruskin's *Seven Lamps of Architecture* as a framework, calls the architect to look back at the previous generations to restore an outlook to architecture which builds structures that are not just utilitarian, but also true and beautiful.

4 || Truth, Beauty, and the Reflection of God: John Ruskin's *Seven Lamps of Architecture* and *The Stones of Venice* as Palimpsests for Contemporary Architecture

MARK R. HALL

Oral Roberts University

Abstract

The guiding lights of modern architecture mainly focus on form and function. However, historically, architecture has been guided by a deeper sense of calling. John Ruskin, a 19th century critic, used the Gothic style of cathedrals as an example to his contemporaries of the transcendental and moral ideals of architecture, which he categorizes as seven lamps or laws. Just as Gothic architecture served as a palimpsest to Ruskin, Ruskin's work is beginning to serve as a palimpsest to a new generation of architects whose designs and structures incorporate various aspects of his seven lamps.

1   Introduction

Architecture is invariably shaped by both its creator and the landscape from which it emerges. These elements are inextricably intertwined to produce a structure that is aesthetically pleasing, philosophically erudite, and fully functional. Nowhere is this more clearly established than with John Ruskin, a noteworthy Victorian art and social critic. His *Seven Lamps of Architecture* and *The Stones of Venice* serve as palimpsests for contemporary architecture. A link to the past is forged based on foundational moral, ethical, philosophical, and religious principles that are reflected in the structures themselves. For Ruskin, when first principles are applied, aesthetic integrity is maintained, truth and beauty are manifested, and the reflection of God is contained in the building itself. The architecture may also point beyond itself to something else, complementing it, expanding it, or transforming it (such as in Gothic architecture). Applying these Ruskinian laws and virtues to today's architecture provides a framework that grounds the discipline in meaningful theological and philosophical underpinnings from which inspiration and creativity may emerge. Contemporary examples include Daniel Libeskind's Jewish Museum Berlin that opened in 2001 and Peter Eisenman's Holocaust Memorial built in Berlin in 2004. Daniel Libeskind is also an architect for the One World Trade Center scheduled to be opened in 2014. The approach taken by these men to design these structures demonstrates their philosophy that architecture should arise out of history and landscape. Therefore, the principles of Ruskin function as a palimpsest for the inspiration, creativity, and designs of architects like Eisenman and Libeskind, as they seek to recapture and maintain the past through structures that promote their own interpretation of memory and beauty, and also reflect truth, power, and life.

2   The Seven Lamps of Architecture

In *The Seven Lamps of Architecture* (1845), John Ruskin defines architecture as "the art which so disposes and adorns the edifices raised by man [. . .] that the sight of them may contribute to his mental health, power, and pleasure"(Ruskin, 1920, p. 8). He asserts that good architecture must exhibit seven lamps that represent spirits or laws: sacrifice, truth, power, beauty, life, memory, and obedience. Ruskin sees these as the framework for architectural creation and design. He believes that good and beautiful architecture must conform to these laws, and the observer should see that "there is room for the marking of his [man's] relations with the mightiest, as well as the fairest, works of God; and that those works themselves have been permitted, by their [the architects'] Master and his [man's], to receive an added glory from their association with earnest efforts of human thought"(Ruskin, 1920, p. 73). According to Ruskin, architecture that reflects these seven lamps will draw the builder and the observer toward an experience with the Master Builder, God.

In *The Seven Lamps of Architecture*, Ruskin explains the meaning of the seven lamps. The illustration below (Figure 4.1) shows the connections that exist among these seven laws (Baljon, 1997, p. 402).

1. Sacrifice – Architecture is an offering to God demonstrating men's "love and obedience and surrender of themselves and theirs to His will"(Ruskin, 1920,

p. 16) as evidenced by the building of beautiful, ornate churches.

2. Truth – Builders must use honest and true materials—crafted by human hands, not machines—respecting them and rejecting false ones.

3. Power – The construction of buildings must focus on mass, quantity of shadow, breadth, sense of surface, size, weight, and shadow; the efforts of the builders through their imagination should point toward the sublimity and majesty of nature.

4. Beauty – Architecture should point individuals toward God and reflect the design and decoration found in nature.

5. Life – Buildings should bear the mark of human hands, celebrating the irregularity in design to show that the ornamentation is not mechanical and demonstrating the joy of the builders as they construct with freedom.

6. Memory – Architecture should respect the social, historical, and cultural character of its milieu, distinguishing between essential and inessential forms.

7. Obedience – Originality must recognize and be restrained by obedience to tradition, especially connecting with the English architecture that has preceded it.

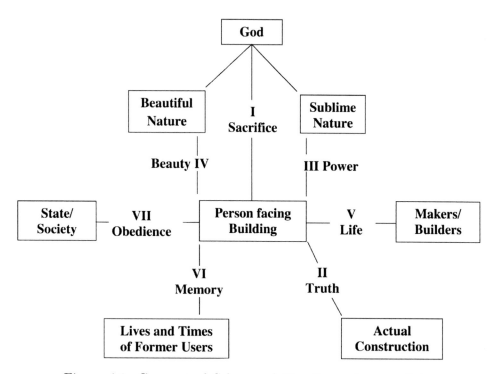

Figure 4.1: Conceptual Scheme of *The Seven Lamps of Architecture*

Ruskin ties beauty to human beings and their experience with nature in *The Seven Lamps of Architecture.* In one diary entry dated April 19, 1846, Ruskin describes a day in Champagnole, France and then comments on how nature affected him:

> I felt it more than usual, but it struck me suddenly how utterly different the impression of such a scene would be, if it were in a strange land and in one without history. How dear to the feeling is the pine of Switzerland compared to that of Canada! I have allowed too little weight to these deep sympathies, for I think, if that pine forest had been among the Alleghanys, or if the stream had been Niagara, I should only have looked at them with intense melancholy and desire for home. (Ruskin, 1956, p. 325)

This observation of creation enables Ruskin to embrace the theory of associationism, especially its connections to history, which influences his aesthetic appreciation. George Landow (1971) points out that Ruskin's emphasis on beauty seems to emerge out of these historical associations that assist his criticism of contemporary architecture. Ruskin finds the homes and public buildings of his England constructed without style, without regard to permanence and without meaning for the men who inhabit them. Since he wishes to correct these deficiencies, he places great emphasis upon historical associations, whose presence, he says, will insure both that an edifice will influence the life of the inhabitant and that it will be solidly constructed — this latter because if a building is to endure long enough for historical associations to accrue, then it must be well made.

Thus, Ruskin's establishment of memory as one of his seven laws—with its focus on the social, historical, and cultural milieu—becomes essential to his philosophy of architecture.

3   Ruskin and *The Stones of Venice*

In *The Stones of Venice,* Ruskin's vivid description of St. Mark's Cathedral (Figure 4.2), a most magnificent structure in Venice—"the most precious building in Europe standing yet in the eyes of men and the sunshine of heaven"(Ruskin on St. Mark's, 1880)—and his detailed sketches of the same (Figures 4.3, 4.4, 4.5, 4.6, 4.7, and 4.8) demonstrate the ability of the author to pen with passion and eloquent style, and the artist to draw with precision and color, the beauty of its architecture:

> A multitude of pillars and white domes, clustered into a long low pyramid of coloured light; a treasure-heap, it seems, partly of gold, and partly of opal and mother-of-pearl, hollowed beneath into five great vaulted porches, ceiled with fair mosaic, and beset with sculpture of alabaster, clear as amber and delicate as ivory —sculpture fantastic and

Figure 4.2: St. Mark's Cathedral, by John W. Bunney—
Public Domain

involved, of palm leaves and lilies, and grapes and pomegranates, and
birds clinging and fluttering among the branches, all twined together
into an endless network of buds and plumes; and, in the midst of it, the
solemn form of angels, sceptred, and robed to the feet, and leaning to
each other across the gates, their figures indistinct among the gleam-
ing of the golden ground through the leaves beside them, interrupted
and dim, like the morning light as it faded back among the branches of
Eden, when first its gates were angel-guarded long ago. And round the
walls of the porches there are set pillars of variegated stones, jasper and
porphyry, and deep-green serpentine spotted with flakes of snow, and
marbles, that half refuse and half yield to the sunshine, Cleopatra-like,
"their bluest veins to kiss"—the shadow, as it steals back from them,
revealing line after line of azure undulation, as a receding tide leaves the
waved sand; their capitals rich with interwoven tracery, rooted knots of
herbage, and drifting leaves of acanthus and vine, and mystical signs,
all beginning and ending in the Cross; and above them, in the broad
archivolts, a continuous chain of language and of life—angels, and the
signs of heaven and the labours of men, each in its appointed season
upon the earth; and above these another range of glittering pinnacles,
mixed with white arches edged with scarlet flowers,—a confusion of delight, amidst which the breasts of the Greek horses are seen blazing in
their breadth of golden strength, and the St. Mark's Lion, lifted on a
blue field covered with stars, until at last, as if in ecstacy, the crests of
the arches break into a marble foam, and toss themselves far into the
blue sky in flashes and wreaths of sculptured spray, as if the breakers
on the Lido shore had been frost-bound before they fell, and the sea-
nymphs had inlaid them with coral and amethyst. (Ruskin, 1885, vol.
2, ch. 4, sec. 14)

Figure 4.3: The South Side of St. Mark's from the Loggia
of the Ducal Palace, Venice, 1851, by John Ruskin—Public
Domain

Figure 4.4: Archivolt in St. Mark's, 1853, by John Ruskin—Public Domain

Figure 4.5: Basket and Lily Capital, St. Mark's Basilica, Venice, 1849–1852, by John Ruskin—Public Domain

Figure 4.6: Northwest Angle of the Façade, St. Mark's
Church, 1851, by John Ruskin—Public Domain

Figure 4.7: North West Porch, St. Mark's, Venice, 1877, by John Ruskin—Public Domain

These words and drawings reflect the magnificence that resonates in the actual architecture, authenticating the lamp of beauty, for Ruskin clearly believes that architecture should reflect the design found in nature and point towards the ultimate Master Builder.

With a philosophy based on aesthetics, place, and history, Ruskin appeals to a moral architecture, encouraging builders to reject the techniques discovered in the Renaissance and developed in the Industrial Revolution and to embrace a time when the best buildings were constructed—the medieval Gothic cathedrals of England and Venice. In his later book, *The Stones of Venice* (1851–1853), Ruskin describes the elements of the Gothic that became foundational for the kind of architecture he proposes, and he provides many examples to illustrate. He points out the three virtues of a building: (1) "That it act well," (2) "That it speak well," and (3) "That it look well" (Ruskin, 1885, vol. 1, ch. 2, sec. 1).

In *The Crown of Wild Olive*, Ruskin explains the purpose of his writing:

> The book I called "The Seven Lamps" was to show that certain right states of temper and moral feeling were the magic powers by which all good architecture, without exception, had been produced. "The Stones of Venice" had, from beginning to end, no other aim than to show that the Gothic architecture of Venice had arisen out of, and indicated in all its features, a state of pure national faith, and of domestic virtue; and that its Renaissance architecture had arisen out of, and in all its features indicated, a state of concealed national infidelity, and of domestic corruption. (Ruskin, 1866, p. 53)

For Ruskin, moral feeling, states of temperament, and architecture cannot be separated. He sees the "moral elements of Gothic" as follows: (1) savageness, (2) changefulness, (3) naturalism, (4) grotesqueness, (5) rigidity, and (6) redundance,

Figure 4.8: Part of St. Mark's, Venice, Sketch After Rain,
1846, by John Ruskin—Public Domain

when "belonging to the building," and (1) savageness or rudeness, (2) love of change,
(3) love of nature, (4) disturbed imagination, (5) obstinancy, and (6) generosity, when
"belonging to the builder" (Ruskin, 1885, p. 155). Thus, Ruskin was not arguing for
a new style of architecture. He was lamenting the plainness and the soullessness
of the architecture designed and built since the Gothic cathedrals of the medieval
period. He "found certain styles (e.g., Baroque) unacceptable because they exploited
illusions, and therefore were not 'truthful'" (Curl, 2006, p. 669). Therefore, according
to Ruskin, in order for architecture to be sincerely honest and truly beautiful, it must
be connected to nature, rooted in right history, and constructed by human hands.

4   Review of Ruskin's Reputation

The appeal of Ruskin's philosophy of architecture was paramount during the Victorian
period. Professor Robert Kerr, a contemporary of the art critic, had previously
espoused the same ideas as Ruskin, but he had left them behind after working twenty

years in the field. He encouraged experienced architects to deter younger apprentices from the idealistic and romanticized views of Ruskin, for Kerr viewed the architect as "a servant of the public for the efficient design of buildings, precisely like the engineer." When he presented a lecture entitled "Architectural Criticism" at the Royal Institute of British Architects, he severely criticized Ruskin saying that "Mr. Ruskin's thoughts soar high enough in the poetry of visionary art, because poetry is his business, but they cannot stoop down to the plain prosaic details of the structuresque, because building is not his business" (Collins, 1998, pp. 259–260). In an October 1849 review of *The Seven Lamps of Architecture* published in the *Journal of Design*, Matthew Digby Wyatt admired "the excellent spirit" that was present in "this thoughtful, eloquent book." However, he quickly points out that Ruskin "either puts his back against [. . .] further development, or would attempt to bring back the world of art to what its course of actions was four centuries ago!" (Mallgrave, 2009, pp. 121, 438).

   Ruskin does not hesitate to move from art critic to social critic, demonstrating how the architecture itself can become a commentary on the denigration, deterioration, and degradation of society. Even as he praises the majesty of St. Mark's in *The Stones of Venice*, he also notes the ironic contrast that takes place in its shadows as the masses ignore its beauty and the poor grovel in their poverty.

> And what effect has this splendor on those who pass beneath it? You may walk from sunrise to sunset, to and fro, before the gateway of St. Mark's, and you will not see an eye lifted to it, nor a countenance brightened by it. Priest and layman, soldier and civilian, rich and poor, pass by it alike regardlessly. Up to the very recesses of the porches, the meanest tradesmen of the city push their counters; nay, the foundations of its pillars are themselves the seats—not "of them that sell doves" for sacrifice, but of the vendors of toys and caricatures. Round the whole square in front of the church there is almost a continuous line of cafés, where the idle Venetians of the middle classes lounge, and read empty journals; in its centre the Austrian bands play during the time of vespers, their martial music jarring with the organ notes,—the march drowning the miserere, and the sullen crowd thickening round them,—a crowd, which, if it had its will, would stiletto every soldier that pipes to it. And in the recesses of the porches, all day long, knots of men of the lowest classes, unemployed and listless, lie basking in the sun like lizards; and unregarded children,—every heavy glance of their young eyes full of desperation and stony depravity, and their throats hoarse with cursing,—gamble, and fight, and snarl, and sleep, hour after hour, clashing their bruised centesimi upon the marble ledges of the church porch. And the images of Christ and His angels look down upon it continually. (Ruskin, 1885, vol. 2, ch. 4, sec. 15)

Ruskin observes that society and architecture are invariably connected.

Figure 4.9: Crystal Palace, Sydenham, by Achille-Louis Martinet—Public Domain

John Matteson makes this observation concerning Ruskin the social critic: "The architecture was sublime; the human activity around it was an obscene mockery. What good was the building if it could not transform the debauched children who cast lots on its very steps? After *The Stones of Venice*, it was no longer enough for Ruskin to criticize art. It was hierarchies of human beings, not structures of wood and stone, that begged most loudly for his attention" (Matteson, 2002, p. 302).

5 Ruskin's Relevance to Contemporary Architecture

Clearly then Ruskin spoke to the Victorian period, but the question inescapably arises, Can the aesthetic and moral philosophies of a Victorian art and social critic be applicable to design and construction today? Is Ruskin relevant to contemporary architecture?

John Matteson discusses this very question. Citing the building of the Crystal Palace (Figures 4.9 and 4.10), whose "prefabricated components heralded a revolution," which was occurring at the same time as the publication of Ruskin's *Stones of Venice*, Matteson asserts that "Ruskin's ideas were already destined for quaintness in the 1850s" (Matteson, 2002, p. 300). He points out some of the difficulties of applying Ruskin's first principles to contemporary architecture:

> Since Ruskin's time, populations have grown and economic systems have expanded with once unimaginable speed. Construction in our time has to be fast. It must be efficient. It must avoid unnecessary expense. If Ruskin foresaw the further mechanization of physical labor, he was at least spared the sadness of seeing how far that mechanization

Figure 4.10: Queen Victoria Opening the 1862 Exhibition (inside view of Crystal Palace), by Joseph Nash—Public Domain

Figure 4.11: Cathedral of St. John the Divine, Wide Angle View—Copyright © 2011 Kirpaks and licensed for reuse under the Creative Commons Attribution-ShareAlike License

would eventually extend. Ruskin also did not anticipate that the alienation that he saw as poisoning the life of the worker might someday encompass not only the process of construction, but also those of conception and design. He could never have imagined on-line catalogs of design components or the idea that an architect might one day resolve decisions of ornamentation, not with painstaking manual drawing or model-building, but with the click of a mouse. Neither could he have expected that modern buildings would often be commissioned and designed, not by individuals at all, but by impersonal organizations. It would have been strange, indeed, for Ruskin to discover the myriad ways in which architecture could divorce itself from the simple human acts of drawing and carving. (Matteson, 2002, p. 300)

Yet Matteson does not completely reject Ruskin's writings about Gothic architecture, citing the construction of St. John the Divine in Manhattan (Figures 4.11 and 4.12) as an exemplar of Ruskinian ideals. In 1972, after no construction had occurred on the building for thirty years, the dean of St. John the Divine proclaimed that it was time to once again begin work and that "the stonework [would] be done by our own unemployed and underemployed neighbors. We will revive the art of stonecraft"(Matteson, 2002, p. 300). Matteson observes that both the process and the product were "profoundly Ruskinian":

> The spirit of the new construction was profoundly Ruskinian: it entrusted a sacred Gothic edifice to hands that would begin the project raw and untutored, in expectation that, as the structure grew and took shape, so, too, would the skills and souls of the workers. That the cathedral actually did become a literal synthesis of stonecutting and soul-making, an exemplar of Ruskin's demand that the work must affirm the passion of the worker, seems to be confirmed in the words of Simon Verity, one of the master carvers employed in the project: "To be a carver, you have to have a passion for it, to love it with all your heart. It's a desire to create order out of chaos, to seek harmonies." (Matteson, 2002, pp. 300–301)

For Matteson, unskilled human hands touching and carving stone so that both are built together reflect the perfect aesthetic and moral for the Ruskinian model, celebrating Ruskin's laws of life and truth. He concludes, "Surely, Ruskin would have applauded this method of construction, a combination, someone has said, of outreach and up-reach. And yet his applause might have been tempered by the knowledge of how deeply the impersonality of technology and profit had insinuated themselves into the building of the cathedral" (Matteson, 2002, p. 301).

Figure 4.12: Cathedral of St. John the Divine, The Western façade, including the Rose Window—Copyright © 2008 William Porto and licensed for reuse under the Creative Commons Attribution-ShareAlike License

6   Architecture as a Palimpsest

During the Victorian era, Thomas Carlyle (1830), like Ruskin, also demanded that attention be given to history. In his essay "On History" (1830), he says that meaning in the present and the future can be known only as the past is studied. He writes: "For though the whole meaning lies far beyond our ken; yet in that complex Manuscript covered over with formless inextricably-entangled unknown characters,—nay which is a *Palimpsest*, and had once prophetic writing, still dimly legible there,—some letters, some words, may be deciphered" (author's emphasis) (Carlyle, 1971, p. 56). Uhlig concurs with Carlyle and maintains that in the intertext, which he likens to the palimpsest (Figures 4.13, 4.14, and 4.15), "historically conditioned tensions come to the fore: tensions not only between calendar time and intraliterary time but also between the author's intention and the relative autonomy of a text, or between the old and the new in general" (Uhlig, 1985, p. 502). The presence of the past coexists with the text; thus, "any text will the more inevitably take on the characteristics of a palimpsest the more openly it allows the voices of the dead to speak, thus—in a literary transcription of our cultural heritage—bringing about a consciousness of the presentness of the past" (Uhlig, 1985, p. 502). Deciphering the present moment of the text as it relates to many past moments reveals the intertextual meaning the text seeks to convey and the critic to uncover.

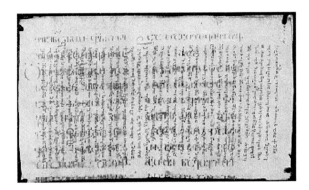

Figure 4.13: A Georgian palimpsest of the 5th/6th century—
Public Domain

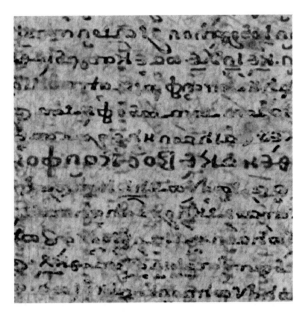

Figure 4.14: Archimedes Palimpsest—Copyright ©
Rochester Institute of Technology, Equipoise Imaging and
Boeing LTS and licensed for reuse under the Creative
Commons Attribution-ShareAlike License

Figure 4.15: Archimedes Palimpsest—Copyright © Rochester Institute of Technology, Equipoise Imaging and Boeing LTS and licensed for reuse under the Creative Commons Attribution-ShareAlike

The word "palimpsest" derives from παλίμψηστος (*palimpsestos*) which is Greek in origin and means "scraped again" (Liddell & Scott, 1990) and can be defined as "a papyrus or other kind of writing material on which two or more sets of writing had been superimposed in such a way that, because of imperfect erasure, some of the earlier text could be read through over-writing"(Darville, 2002, p. 309). When used in the field of archaeology, "the term is often applied to landscapes in which traces of earlier arrangements can be seen amongst and below the modern pattern"(Darville, 2002, p. 309), and in architecture palimpsest means the shadow of a past structure that is in some way incorporated as part of an old one that has been remodeled or a new one that has been built. Michael Earle describes the concept as follows:

> Architects use the concept of palimpsest to imply a ghost, an image of what once was. Of course, in the built environment, this occurs often, whenever spaces are shuffled, rebuilt, or remodeled, shadows remain. Tarred rooflines remain on the sides of a building long after the neighboring structure has been demolished and long ago removed stairs leave a mark where the painted wall surface stopped. Dust lines remain from a relocated appliance. Ancient ruins speak volumes of their former wholeness. Palimpsests can inform us of the realities of the built past. (Earle, 2012)

According to Peter Eisenman, an architect and theorist, the palimpsestic connection of site history with contemporary design and construction is essential: "Any site contains not only presences, but the memory of previous presences and the immanences of a possible presence. The physical difference between a moving thing (dynamism) and a still one (stasis) is that the moving one contains the trace of where it has been and where it is going." He then connects the history to the city itself, seeing it as an integral part of the site: "The introduction of this trace, or condition of absence, acknowledges the dynamic reality of the living city" (Eisenman, 2004, p. 207). Eisenman describes an architectural palimpsestic text as follows:

> In my proposal for rhetorical figures, architecture is no longer elements but an *other* grammatical counter, proposing an alternate reading of the idea of site and object. In this sense, a rhetorical figure will be seen to be inherently contextual in that the site is treated as a deeply scored palimpsest. [. . .] This text suggests that there are other meanings which are site specific by virtue of their pre-existence, however latent within the context. (Eisenman, 2004, p. 206)

He explains that the word "text" when used in relationship to architecture

> can be used for any and all strategies and conditions which dislocate architecture from its authorial or natural condition of being; that is, the detaching of what architecture looks like from the need to represent function, shelter, meaning and so forth. It is not so much that the look of architecture will change (architecture will always look like architecture) but rather the style and significance of its look will be different. The idea of text is not in opposition to the reality of architecture, just as the imaginary is not the opposite of the real; it is an *other* discourse. Text surrounds reality at the same time that it is internal to reality. (Eisenman, 1988)

Eisenman, like Ruskin, sees that architecture communicates a text beyond its outward beauty: "Thus in architecture it is possible to say that text is what always exceeds the immediate response to a visual or sensory image, i.e. that which we see on the surface as the story, or that which we see as the beautiful. This is the heart of the matter"(Eisenman, 1988). Thus, a palimpsest can be defined as that text which underlies another text (an ur-text)—a present text with origins in a past one (palingenesis) or at least shaped by an underlying one (ananke)—or a text that influences something not of its own genre—art, music, architecture (Uhlig, 1985, p. 503).

7   Peter Eisenman and *The Memorial for the Murdered Jews of Europe*

Finished in 2004 and inaugurated on May 10, 2005—sixty years after the conclusion of World War II—The Memorial to the Murdered Jews of Europe (Figures 4.16 and 4.17), also known as the Holocaust Memorial, was built in Berlin by Peter Eisenman, an American architect (Brunberg, 2009). Encompassing five and a half acres

Figure 4.16: Memorial to the Murdered Jews of Europe, Peter Eisenman—Public Domain

Figure 4.17: Memorial to the Murdered Jews of Europe, Peter
Eisenman—Copyright © 2005 de:Benutzer:Schreibkraft and
licensed for reuse under the Creative Commons Attribution-
ShareAlike License

(Ouroussoff, 2005), it is designed with "2,711 pillars, planted close together in un-
dulating waves, represent[ing] the 6 million murdered Jews" (Quigley, 2005). The
memorial is open every day year round and can be entered on each of the four sides
(Quigley, 2005).

True to his architectural theory, Eisenman is focused on incorporating the
memorial into its site and to the city itself (Quigley, 2005), "acknowledge[ing] the
dynamic reality of the living city" (Eisenman, 2004, p. 207). Nicolai Ouroussoff
explains:

> At first, you retain glimpses of the city. The rows of pillars frame a
> distant view of the Reichstag's skeletal glass dome. To the west, you
> can glimpse the canopy of trees in the Tiergarten. Then as you descend
> further, the views begin to disappear. The sound of gravel crunching
> under your feet gets more perceptible; the gray pillars, their towering
> forms tilting unsteadily, become more menacing and oppressive. The
> effect is intentionally disorienting. (Ouroussoff, 2005)

The construction arises out of the city's history, bringing it into the present:
"The memorial's grid, for example, can be read as both an extension of the streets that
surround the site and an unnerving evocation of the rigid discipline and bureaucratic
order that kept the killing machine grinding along. The pillars, meanwhile, are an
obvious reference to tombstones" (Ouroussoff, 2005). Yet observation alone is not
enough; one must experience the site "as a physical space" in order to truly understand
it:

> No clear line, for example, divides the site from the city around it. The
> pillars along its periphery are roughly the height of park benches. A few
> scattered linden trees sprout between the pillars along the memorial's
> western edge; at other points, outlines of pillars are etched onto the
> sidewalk, so that pedestrians can actually step on them as they walk
> by. (Ouroussoff, 2005)

Sarah Quigley, a novelist and critic, describes her encounter with the memorial:

> Even on bright sunny days, the stones look sober and drab. Standing on an uneven piece of land, the stelae almost fall into the centre of the site, rising up again towards the edge, forming a myriad of uneven stone corridors. Walking down one of these passages is disorientating, and scary; you can't see who is approaching you, nor who is behind. The tilting ground and lack of vision offers some small idea of the Jewish experience from WWII: your past snatched away, your future insecure, little hope of escape. (Quigley, 2005)

In this memorial the past haunts both the present and the future.

Somewhat unexpectedly, Eisenman rediscovered his Jewishness in this architecture: "[With this work] I came back to the heart of my identity" (Quigley, 2005). Even so, Eisenman is not interested in viewing the Holocaust with sentimentality. He does not "want people to weep and then walk away with a clear conscience" (Ouroussoff, 2005). He wants all who visit to realize their culpability, to understand "the process that allows human beings to accept such evil as part of the normal world - the incremental decisions that collectively lead to the most murderous acts" (Ouroussoff, 2005). Eisenman "leaves you standing on the edge of the abyss. In so doing, he suggests that the parameters of guilt are not so easily defined: it includes those who looked the other way, continued with their work, refused to bear witness. It is true of Americans as well as Germans, Roman Catholic clerics as well as Nazi secretaries" (Ouroussoff, 2005).

In contrast to Ruskin who believed that architecture should reflect beauty and point upward to the ultimate Maker, Eisenman's design is plain and its purpose is to cause the viewer to look inward. Although Paul Spiegel, a leader of the Jews in Germany, felt that the memorial was "incomplete" because it did not shock those who saw it with its history, Eisenman's desire was to promote and elicit a response that concerned more than just the Holocaust; he wanted people to focus on anti-Semitism in general and civilization's response to it. This discussion would broaden the appeal of the memorial and make it a part of the daily life of the city (Quigley, 2005). Perhaps this statement encapsulates Eisenman's attitude most of all: "I think people will eat their lunch on the pillars. [. . .] I'm sure skateboarders will use it. People will dance on top of the pillars. All kinds of unexpected things are going to happen" (Quigley, 2005). Eisenman's prediction has already come true, for Nicolai Ouroussoff writes, "The day I visited the site, a 2-year-old boy was playing atop the pillars - trying to climb from one to the next as his mother calmly gripped his hand" (Ouroussoff, 2005).

The palimpsest of the Holocaust surrounds the site. Nicolai Ouroussoff asserts, "The location could not be more apt. During the war, this was the administrative locus of Hitler's killing machine. His chancellery building, designed by Albert Speer and since demolished, was a few hundred yards away just to the south; his bunker

Figure 4.18: The Jewish Museum Berlin, to the left of the old Kollegienhaus. Designed by Daniel Libeskind—Copyright ©️ 2008 Daniel Libeskind and licensed for reuse under the Creative Commons Attribution-ShareAlike License

lies beneath a nearby parking lot" (Ouroussoff, 2005). Although criticized by some well-known Germans for its abstract symbolism, its dreary atmosphere, and its sparse construction (Quigley, 2005) (e.g., no names are etched into the pillars [Brunberg, 2009]), Eisenman insists that The Memorial for the Murdered Jews of Europe "is both perfect in its symbolism, and a necessary aid to atonement. 'It stands there, silent,' he says: 'the one who has to talk is you'" (Quigley, 2005).

8   Daniel Libeskind and His Architecture

8.1   *The Jewish Museum Berlin*

Opened in 2001, The Jewish Museum Berlin (Figure 4.18) showcases 1700 years of the history of the Jews in Germany. Two buildings house the exhibits, the old *Kollegienhaus*, once used as a courthouse, and a new one designed by Daniel Libeskind. The museum covers 166,840 square feet (Libeskind, 2011) and is constructed as a twisted zig-zag to remind museum-goers of a warped Star of David (Mueller-Kroll, 2011). It is entered through an underground tunnel. A "Void"—a space with nothing in it except 10,000 iron faces that are called "Fallen Leaves," created by an artist from Israel, Menashe Kadishman—is part of the memorial (Installations, 2012). One visitor describes his experience in this manner:

> On the floor, thousands of pieces of heavy metal cut into shapes of the faces of screaming holocaust victims. The visitor is encouraged to walk across the void. Clank, clank, clank echoing up into and all around the void. The noise rings in your head but there is no escape because as you are tempted to look down the screaming faces stare into your psyche. Very simple, very effective. Haunting. (Gold, 2004)

The memorial has three intersecting tunnels that are said to represent three pathways of German life for the Jew: the Axis of Continuity (with German history), the Axis of Emigration (from Germany), and the Axis of the Holocaust. Then the participant moves into the Garden of Exile with its 49 pillars that reminds visitors of the people expelled from Germany, which according to Libeskind, is designed "to completely disorient the visitor. It represents a shipwreck of history." Even so, Russian willow oak trees that represent hope have been planted on top of the stelae (Libeskind Building, 2012).

Libeskind's design entitled "Between the Lines" was chosen from a world-wide competition of 165 entries (Levenson, 2005), and, of course, the architect was ecstatic when he won: "It was a thrilling moment when I was selected. The jury recognized that my plan was neither dogmatic nor glib; that it served as an individualized mirror, which each visitor could read in a different way. They valued its authenticity and celebrated its originality. I felt honored and elated" (Libeskind, 2004, p. 85).

Because of his own personal background and experience, Daniel Libeskind knew that the architecture must first connect the place to its history and then take visitors from the past to the present and propel them to the future, experiencing a sense of alienation:

> You struggle to find the most immediate way to get at the truth. What was needed, as I saw it, was a building that, using the language of architecture, speaking from its stones, could take us all, Jews and non-Jews alike, to the crossroads of history, and show us that when the Jews were exiled from Berlin, at that moment, Berlin was exiled from its past, its present, and—until this tragic relationship is resolved— its future. (Libeskind, 2004, p. 83)

At this museum, Daniel Libeskind believes history and architecture are joined, for this place "thematizes and integrates, for the first time in post-war Germany, the history of the Jews in Germany, the repercussions of the Holocaust and spiritual displacement. It is also just a museum with exhibits on the wall" (Mueller-Kroll, 2011).

8.2   *The One World Trade Center*

Winning the design competition in 2003 out of 13,683 entrants with his Memory Foundations plan (titled this, per Libeskind, "because it's about memory and at the center of it is a foundation for 21st century New York" [Nessen, 2011])—originally known as the Gardens of the World (Hirschkorn, 2003; Swanson, 2011; NY1 News, 2003), Daniel Libeskind was chosen as the architect to create the Ground Zero Master Plan for the reconstruction of the World Trade Center (Figure 4.19) (Libeskind, 2011). As he put together the design, he realized, "We have to be able to enter this hallowed, sacred ground while creating a quiet, meditative and spiritual space" (Studio

Figure 4.19: Ground Zero Master Plan (2006)—Copyright ©
Silverstein Properties

Libeskind, 2012). He was very sensitive to the site and to New Yorkers, desiring for his plan to fully memorialize what had happened there:

> When I first began this project, New Yorkers were divided as to whether to keep the site of the World Trade Center empty or to fill the site completely and build upon it. I meditated many days on this seemingly impossible dichotomy. To acknowledge the terrible deaths which occurred on this site, while looking to the future with hope, seemed like two moments which could not be joined. I sought to find a solution which would bring these seemingly contradictory viewpoints into an unexpected unity. So, I went to look at the site, to stand within it, to see people walking around it, to feel its power and to listen to its voices. (Studio Libeskind, 2012)

For Libeskind, this project was personal: "What happened on 9/11 was not something abstract, it happened to me" (qtd. in Earle). In fact, on the day Libenskind opened his Jewish Museum in Berlin, the Twin Towers in New York were attacked and then collapsed. As soon as he received word around 2:30 p.m., he left for the States. He still remembers that day, "I turned to all my colleagues [. . .] and I do not know where it came from, but I said, 'I'm returning to Lower Manhattan'" (Needham, 2011).

Because of disagreements among all those involved, the project was eventually removed from Libenskind (Needham, 2011). Even though many architectural changes were made, the WTC Masterplan (Figure 4.19) as delineated by Libeskind was still basically followed:

> The WTC Masterplan serves as both the conceptual basis and the technical foundation for the entire complex re-development of ground zero. The Masterplan defines the spirit of the approach to re-building and creates a meaningful conceptual framework for the site. It also defines the spatial organization of all elements of the development within the site with an emphasis on the human experience and the public realm. The Masterplan dictates the location and massing of each program element, building height and relative size, as well as proximity and relationship to one another. The WTC Masterplan also supplies the framework for the site's infrastructure, transportation, sustainability standards and security strategy and lays out the functional relationship between all the site elements with respect to the surrounding context of the immediate neighbourhoods and the surrounding city. (Libeskind, 2011)

Michael Arad, the final designer, credits Libeskind as the one who "'established the broad parameters' of what is now the new World Trade Center and 'acted as a guidestar. If you're going to build something, you need to start some place.'"

Libeskind acknowledges his part in the process: "I'm so happy to be able to design a piece of this city." He observes, "If you're a conductor or a composer, Stravinsky or Copland, and the New York Philharmonic is performing your piece and you're conducting it, do you regret that you're not playing the first violin? That you're not playing the tuba? Of course not" (Needham, 2011). Therefore, he asserts confidently, "In the end, the public will see the symbolism of the site. [. . .] Of course, compromises had to be made, but a master plan is not about a few lines drawn on paper. It's about an idea, and how to express that idea through the turmoil of politics and the creativity of all the other architects. In the end, the result will be pretty close to my original rendering" (Davidson, 2007).

Libenskind's original plan reflects his intense interest in symbolism. He wanted the foundations of the former buildings to be part of the memorial site ("We need to journey down, some 70 feet into Ground Zero, onto the bedrock foundation, a procession with deliberation into the deep indelible footprints of Tower One and Tower Two"), and he emphasized their connection to the nation itself.

> The great slurry walls are the most dramatic elements which survived
> the attack, an engineering wonder constructed on bedrock foundations
> and designed to hold back the Hudson River. The foundations with-
> stood the unimaginable trauma of the destruction and stand as eloquent
> as the Constitution itself asserting the durability of Democracy and the
> value of individual life. (Studio Libeskind, 2012)

Libeskind imagined "the sky" as "home again" to "vertical gardens" on "a towering spire of 1776 feet high" (symbolic of the founding of the country, the year when the Declaration of Independence was signed)—the "Gardens of the World," filled with plants from all parts of the earth (Studio Libeskind, 2012; NY1 News, 2003; Nessen, 2011). He explains, "Why gardens? Because gardens are a constant affirmation of life. A skyscraper rises above its predecessors, reasserting the preeminence of freedom and beauty, restoring the spiritual peak to the city, creating an icon that speaks of our vitality in the face of danger and our optimism in the aftermath of tragedy" (Studio Libeskind, 2012).

Reminiscent of the Statue of Liberty, the tower would be off-center in its northwest corner, designed to pay homage to the Statue of Liberty's torch which Libeskind remembers seeing when he was 13 years old in 1959 when he came to the United States from Poland (Swanson, 2011). Indeed Libeskind's ideas emerge out of his experience as an immigrant. He explains in his proposal for the reconstruction of Ground Zero: "I arrived by ship to New York as a teenager, an immigrant, and like millions of others before me, my first sight was the Statue of Liberty and the amazing skyline of Manhattan. I have never forgotten that sight or what it stands for. This is what this project is all about" (Studio Libeskind, 2012).

The Wedge of Light piazza and the Park of Heroes open spaces were significant places in Daniel Libeskind's plan (Lower Manhattan Development Corporation, 2003).

Libeskind explains how his design remembers the ones who died: "Those who were lost have become heroes. To commemorate those lost lives, I created two large public places, the Park of Heroes and the Wedge of Light. Each year on September 11th between the hours of 8:46 a.m., when the first airplane hit and 10:28 a.m., when the second tower collapsed, the sun will shine without shadow, in perpetual tribute to altruism and courage" (Studio Libeskind, 2012). Once again, the symbolism is paramount.

The construction of the lynchpin building finally started in 2006 and is scheduled to be finished in 2014. The One World Trade Center, or the 1 WTC, previously called the Freedom Tower (Figure 4.20), will occupy the place where the original 6 World Trade Center stood. When completed, the 1 WTC will be the tallest building in the Western Hemisphere rising 1,776 feet as originally envisioned by Libeskind (Council on Tall Buildings and Urban Habitat, 2012). Like Ruskin and Eisenman, Libeskind's design is inextricably linked to history. As Michael Earle observes, "In terms of design, his best buildings are connected strongly to history and are deeply influenced by it." His masterplan is "a palimpsest of the site itself" (Earle, 2012). The past coexists with the architectural texts, and thus reaffirms that "any text will the more inevitably take on the characteristics of a palimpsest the more openly it allows the voices of the dead to speak, thus [. . .] bringing about a consciousness of the presentness of the past" (Uhlig, 1985, p. 502). Earle acknowledges the changes made to the masterplan but affirms its influence: "While some other parts of the masterplan have been eliminated or changed in political wrangling, the design remains true to itself. As I write this, we are 4 days from the 10th anniversary of September 11th 2001 and the plan that Libeskind created has enough remaining power to make the place where so many people perished, a historical site whose architecture proudly defends its memories" (Earle, 2012). As demonstrated through his symbolism, the design has been connected to memory, one of the seven laws of architecture delineated by Ruskin, as he affirms that architecture must respect the social, historical, and cultural character of its surroundings. Earle concludes, "The design stands as a true description of palimpsest. As this important anniversary comes and goes, we can appreciate the work of great architecture and design which helps to commemorate that awful moment when the world changed forever" (Earle, 2012). The One World Trade Center stands—arising from the palimpsest of September 11, 2001—and reflects both the tragedy and the triumph of the site.

In architecture, site and design are inseparably linked to produce a structure that focuses on the lamps of truth, power, beauty, life, and memory, as delineated by John Ruskin. These ideals have in some profound ways become the palimpsest for contemporary architects, such as Peter Eisenman and Daniel Libeskind, demonstrating that "any site contains not only presences, but the memory of previous presences and the immanences of a possible presence" (Eisenman, 2004, p. 207). In these structures built to commemorate the Holocaust and the tragedy of 9-11, history haunts the visitors—the past informs the present that prepares the participants for the future.

Figure 4.20: One World Trade Center design released in May 2012—Public Domain

They experience the horrible events that happened there and are forced to embrace what lies ahead.

References

Baljon, C. J. (1997). Interpreting Ruskin: The argument of *The Seven Lamps of Architecture* and *The Stones of Venice*. *The Journal of Aesthetics and Art Criticism*, 55(5), 401–414.

Brunberg, J. (2009). Memorial to the murdered Jews of Europe. *The Polynational War Memorial*, August 31. Available from http://archive.is/gEP6B

Carlyle, T. (1971). On history. In *Thomas carlyle: Selected writings* (pp. 51–58). New York: Penguin Books.

Collins, P. (1998). *Changing ideals in modern architecture, 1750–1950.* Montreal: McGill-Queen P, 2nd edition.

Council on Tall Buildings and Urban Habitat (2012). One World Trade Center. Available from http://www.skyscrapercenter.com/new-york-city/one-world-trade-center/

Curl, J. S. (2006). Ruskin, John. In *A dictionary of architecture and landscape architecture*. New York: Oxford University Press.

Darville, T. (2002). Palimpsest. In *The concise Oxford dictionary of archaeology* (p. 309). New York: Oxford University Press.

Davidson, J. (2007). The liberation of Daniel Libeskind. *New York Magazine*, September 30, 56–64. Available from http://nymag.com/arts/architecture/features/38356/

Earle, M. (2012). Daniel Libeskind: The architecture of palimpsest. *Think Design Magazine*. Available from http://thinkdesignmagazine.com/index.php/architecture/daniel-libeskind

Eisenman, P. (1988). Architecture as a second language: The texts of between. *Threshold, the Journal of the School of Architecture*, 4, 72.

Eisenman, P. (2004). Architecture and the problem of the rhetorical figure. In *Eisenman inside out: Selected writings, 1963–1988* (pp. 202–207). New Haven: Yale University Press.

Gold, M. (2004). German jewry's tragic history: Melvin Gold visits Berlin's Jewish museum. *Chigshul Magazine*, September 4. Available from http://www.jmberlin.de/main/DE/06-Presse/02-Pressespiegel/artikel/2004/2004_09_04_cm.php

Hirschkorn, P. (2003). 9/11 memorial design contest called biggest ever. *CNN U.S.*, May 30. Available from http://edition.cnn.com/2003/US/Northeast/05/30/ wtc.memorial/

Installations (2012). The installations. *The Jewish Museum Berlin (jmberlin.de)*. Available from http://www.jmberlin.de/main/EN/01-Exhibitions/04-installations.php

Landow, G. P. (1971). Ruskin's refutation of "false opinions held concerning beauty". In *The aesthetic and critical theories of John Ruskin* Princeton: Princeton University Press. Available from http://www.victorianweb.org/authors/ruskin/ atheories/2.1.html

Levenson, G. (2005). The captivating Jewish museum. *The Jewish Week*. Available from http://www.highbeam.com/doc/1P1-108595819.html

Libeskind, D. (2004). *Breaking ground: Adventures in life and architecture.* East Rutherford, NJ: Penguin Putnam.

Libeskind, D. (2011). Ground zero master plan: New York, NY. *Studio Daniel Libeskind (daniel-libeskind.com)*. Available from http://daniel-libeskind.com/ projects/ground-zero-master-plan

Libeskind Building (2012). The Libeskind Building. *The Jewish Museum Berlin.* Available from http://www.jmberlin.de/main/EN/04-About-The-Museum/ 01-Architecture/01-libeskind-Building.php

Liddell, H. G. & Scott, R. (1990). *A Greek-English lexicon.* New York: Clarendon Press, 9th edition.

Lower Manhattan Development Corporation (2003). Lower Manhattan transportation strategies. Available from http://www.renewnyc.com/plan_des_dev/ transportation/

Mallgrave, H. F. (2009). *Modern architectural theory: A historical survey, 1673–1968.* New York: Cambridge University Press.

Matteson, J. (2002). Constructing ethics and the ethics of the construction: John Ruskin and the humanity of the builder. *Crosscurrents*, Fall 2002, 294–303. Available from http://www.questia.com/library/1G1-94983815/constructing-ethics-and-the-ethics-of-construction

Mueller-Kroll, M. (2011). Jewish Museum Berlin celebrates 10th anniversary. Available from http://www.npr.org/2011/10/19/141521740/jewish-museum-berlin-celebrates-10th-anniversary

Needham, P. (2011). Daniel Libeskind: The return of ground zero's master planner. *Huffington Post.* Available from http://www.huffingtonpost.com/2011/09/09/daniel-libeskind-ground-zero_n_954949.html

Nessen, S. (2011). Q&A: Interview with World Trade Center site architect Daniel Libeskind. Available from http://www.wnyc.org/articles/wnyc-news/2011/sep/07/interview-master-architect-world-trade-center-site-daniel-libeskind/

NY1 News (2003). Libeskind plan chosen for WTC site. Available from http://www.ny1.com/content/news/28221/libeskind-plan-chosen-for-wtc-site-

Ouroussoff, N. (2005). A forest of pillars, recalling the unimaginable. *The New York Times*, May 9. Available from http://www.nytimes.com/2005/05/09/arts/design/09holo.html?pagewanted=all&_r=0

Quigley, S. (2005). Holocaust Memorial: Architect Peter Eisenman, Berlin 2005. *The Polynational War Memorial.* Available from http://www.war-memorial.net/Holocaust-Memorial--Architect-Peter-Eisenman,-Berlin-2005-2.66

Ruskin, J. (1866). *The crown of wild olive.* New York: Colonial Press. Available from http://www.gutenberg.org/ebooks/26716

Ruskin, J. (1885). *The stones of Venice, volumes 1–2.* New York: John B. Alden.

Ruskin, J. (1920). *The seven lamps of architecture.* London: Waverley Book Company. Available from http://openlibrary.org/books/OL13514059M/The_seven_lamps_of_architecture.

Ruskin, J. (1956). *The diaries of John Ruskin: 1835–1847*, volume 1. Oxford: Clarendon Press.

Ruskin on St. Mark's (1880). Ruskin on St. Mark's: His work in connection with the famous Basilica. *The New York Times.* Available from http://query.nytimes.com/mem/archive-free/pdf?res=F10F17FB35551B7A93CBA91789D85F448884F9

Studio Libeskind (2012). New World Trade Center site designs: Firm D: Introduction. Available from http://www.renewnyc.com/plan_des_dev/wtc_site/new_design_plans/firm_d/default.asp

Swanson, C. (2011). Libeskind, Daniel: The lessons of the master planner. *New York Magazine*, August 27. Available from http://www.nymag.com/news/9-11/10th-anniversary/daniel-libeskind/

Uhlig, C. (1985). Literature as textual palingenesis: On some principles of literary history. *New Literary History*, 16, 481–513.

Part III

Software Engineering and Human Agency

Software engineering is rarely considered in connection to philosophy. However, the entire enterprise was conceived by Gödel, Turing, Church, and others in order to answer deep questions about the limitations of computation. This part begins with two papers by Jonathan Bartlett. The first paper discusses how computation can be used as a guidepost to separate out material from non-material causation. The second paper utilizes this concept to outline an improved software complexity metric. The next paper by Winston Ewert et al. refines previous attempts to quantify information by showing how contextual data can affect information measurements and defines a new information metric to take this into account. Finally, a paper by Eric Holloway describes an attempt to directly observe and measure the informational contribution of intelligent beings to search algorithms. While Holloway's attempt was unsuccessful, the experimental setup is unique and promising and will hopefully inspire future researchers to build on his approach.

Using Turing Oracles in Cognitive Models of Problem-Solving

JONATHAN BARTLETT

The Blyth Institute

Abstract

At the core of engineering is human problem-solving. Creating a cognitive model of the task of problem-solving is helpful for planning and organizing engineering tasks. One possibility rarely considered in modeling cognitive processes is the use of Turing Oracles. Copeland (1998) put forth the possibility that the mind could be viewed as an oracle machine, but he never applied that idea practically. Oracles enable the modeling of processes in the mind which are not computationally based. Using oracles resolves many of the surprising results of computational problem-solving which arise as a result of the Tractable Cognition Thesis and similar mechanistic models of the mind. However, as research into the use of Turing Oracles in problem-solving is new, there are many methodological issues.

1 Broad Views of Cognition and Their Historic Consequences in Cognitive Modeling

In the philosophy of mind, three main overarching theories exist concerning how the mind works—physicalism, dualism, and emergentism. These are ways of understanding cognitive processes in their broadest view. Physicalism is the idea that there is nothing going on in the mind that is not describable through standard physical processes. There may yet be physical processes not currently understood or even known, but, in the long run, there should not be anything involved in causal processes that is not physical and understandable through physics. Some physicalists allow for the

non-reduction of mental states to physical states, or at least an epistemological reduction, but they are all clear in the closed causality of the physical (Horgan, 1994).

Dualism is the primary contender for this area. Dualism is the idea that the
mind and the body are not equivalent—that there exists at least some part of human
cognition that is beyond what is describable by physics or using physical entities. It
holds that a reduction of the mind to brain physics does not capture the entirety of
what the mind is doing. It also says that there is a causal element being left out—that
the mind, while not itself entirely physical, participates in the causal chain of human
action. In other words, it is not a purely passive element but has a causal role (Heart,
1994).

A third theory is emergentism. Emergentism tries to split the line between
physicalism and dualism. However, it is a very fluid term and is difficult to distinctly
identify. Some forms of emergentism (especially "weak emergence" or "epistemological emergence") are essentially physicalism, while others (for instance, "strong
emergence" or "ontological emergence") propose outside laws of emergence which
transform the character of initial properties in certain configurations (O'Connor &
Wong, 2012). Therefore, strong emergence tends to be essentially a form of dualism
except that the dualistic properties are captured in a set of laws of emergence. The
question then is whether these laws themselves can be considered material. Since the
question posed in this study concerns whether or not physical causation is sufficient
for an explanation, most views of emergence can be classified as either physicalist or
dualist.

Historically, aspects of cognition that were considered to be part of the non-
physical mind were left unmodeled by dualists. By contrast, the goal of physicalism
is to force all phenomena into explicitly physical models, a process not easily accomplished.

To begin with, defining physicalism and dualism are not easy tasks. For a
dualist to say that there is more than one mode of causation, at least one of those
modes needs to be clearly and explicitly described. Similarly, if a physicalist says that
all causes are physical, such a statement is meaningless without a solid definition of
what counts as physical and what does not (Stoljar, 2009).

Several insufficient definitions of physicalism are often suggested. For example,
one definition is that physicalism deals only with material causes. However, no clear
explanation is given as to what counts as a material cause. Another definition is that
physicalism deals only with observable phenomena. This could have two meanings,
both of which are problematic. If it means that it deals only with things which
can be observed directly, then this would remove most of modern physics from the
physical—direct observation is not possible for atoms, molecules, forces, and the
like. If, on the other hand, the definition includes indirect observations, then there
is no reason to suppose that only physical entities are observable. It is precisely
the contention of the dualists that there are non-physical modes of causation which
have real effects in the world. If dualism is true, then non-physical causes should

be indirectly observable. Therefore, observability can't be a distinguishing factor. A third definition is that physical things are testable. However, this fails for the same reason that the observable definition fails. Testing simply means looking at observations, and determining whether or not they match the expectations of the theory. Therefore, any observable phenomena should be testable in the same way.

One distinguishing factor proposed by physicalists to distinguish between physical and non-physical behavior is computability. With computability, physical processes are those whose results can (at least in theory) be calculated by computational systems, while non-physical processes are those which cannot. This has been proposed by Iris van Rooij in his *Tractable Cognition Thesis* as well as Stephen Wolfram in his *Principle of Computational Equivalence* (van Rooij, 2008; Wolfram, 2002). By using this well-defined theory of computability and incomputability, developed in the early 20th century by Gödel, Church, Turing, and others, it becomes at least possible to make meaningful statements about physical and non-physical processes. In addition, because of the groundwork laid by the same pioneers of computability and incomputability, further advances can be made beyond previous dualist conceptions of the mind which actually include non-physical elements in models of cognition.

2   A Primer on Computability and Incomputability

Incomputability generally refers to the question of whether or not a given function can be computed given a set of operators. So, for instance, given only the addition, subtraction, and summation operators, division cannot be computed. However, given those same operators, a multiplication function can be computed.

One place where incomputability reigns is on self-referential questions. There are numerous questions that can be asked about a set of mathematical operators which cannot be answered solely by the functions of the operators themselves. For example, let's say you have a set of operators (O) and a list (L) of all of the valid functions that take a single value as a parameter, yield a single value as a result, are of finite length, and can be defined using the operators in O, listed in alphabetic order. This is a countable infinity because each function in L can be identified by an ordinal number, which is its index into the list. Now, because all of L are valid functions, there exists a function $F(x)$ which takes the function from L at index x and yields the value of that function with x as the parameter. The question is, is $F(x)$ in L?

The answer, perhaps surprisingly, is no. This means that defining $F(x)$ will require operators not in O. Another example will help demonstrate why. Take another function, $G(x)$, which returns $F(x) + 1$. If $F(x)$ is in L, then, $G(x)$ is also in L (assuming that the addition operator is in O). If $G(x)$ is at index n of L and has a result r when computed using its own index (which is defined as n), by definition, since n is the index of G, then $F(n)$ must return the same result as $G(n)$, which we have defined to be r. However, the definition of $G(x)$ says that it must be $F(x) + 1$! Since $F(n)$ returns r and $G(n)$ returns r, this leads to a contradiction, because r cannot be equal to $r + 1$. Therefore, $F(x)$ cannot be computed using the operators in O. This proof is quite independent of what operators exist in O, provided they are singly valued and include the addition operator. Thus, $F(x)$ is a true function of x but is incomputable with fixed-length programs of operators in O.

Using the example given above, it seems that computability questions are based on the set of operators being used to define it. This is largely true. So, if computability is operator-dependent, how can it help answer questions about the physicality of the process? The Church-Turing thesis provides a solution, stating that all finitary mathematical systems are computationally equivalent to some Turing machine (Turing, 1937, 1939).[1]

Figure 5.1: An example of a working Turing machine, constructed by Mike Davey—Copyright ©2012 Rocky Acosta and licensed for reuse under the Created Commons Attribution License

---

[1] Turing machines are important because their functions can be explicitly described and their operations can be concretely implemented in the real world using machines, and as such they are both verifiable and unambiguous. A Turing machine consists of four parts—an (theoretically) infinitely long tape (i.e., memory), a read/write head for the tape, a state register, and a fixed state transition table. The only unimplementable feature of Turing machines is the requirement for an infinitely long tape. However, in the absence of an infinite tape, it can at least be detected when a given process requires more tape than actually available. One of the purposes of Turing machines was to make explicit what was meant by the terms "algorithm," "effectively calculable," and "mechanical procedure." In other words, the original purpose of developing Turing machines was to delineate between what was calculable and what was not.

The Church-Turing thesis was discovered when several finitary logic systems were developed independently, including Church's lambda calculus (Church, 1936; Turing, 1937). It is hard to imagine two systems so different in approach as Church's lambda calculus and the Turing machine. Yet, in the end, it was proven that they have the exact same computational abilities. To be technically correct, especially with Turing machines, it is their maximum abilities which are equivalent. A Turing machine can be defined with equivalent or less computational power than the lambda calculus, but not with more. Thus, the computational power of finitary systems do imply a fixed set of operators.

Such finitary systems which have this maximal computational power are known as universal machines, or universal computation systems, since they can be programmed to perform any calculation that is possible on a finitary computation system. Thus, any computability question that would be true for one of them would be true for all of them. Therefore, when used without qualification, incomputability usually refers to something which is incomputable on a universal computation system.

Wolfram and van Rooij both use universal computation to set a maximal level of sophistication available in nature. Wolfram explains his Principle of Computational Equivalence:

> One might have assumed that among different processes there would be a vast range of different levels of computational sophistication. But the remarkable assertion that the Principle of Computational Equivalence makes is that in practice this is not the case, and that instead there is essentially just one highest level of computational sophistication, and this is achieved by almost all processes that do not seem obviously simple . . . For the essence of this phenomenon is that it is possible to construct universal systems that can perform essentially any computation—and which must therefore all in a sense be capable of exhibiting the highest level of computational sophistication (Wolfram, 2002, p. 717).

Wolfram is thus stating that within nature, computability is the limiting factor of what is possible. Van Rooij, while restricting his comments to the nature of the mind, makes basically the same point:

> Human cognitive capacities are constrained by computational tractability. This thesis, if true, serves cognitive psychology by constraining the space of computational-level theories of cognition. (van Rooij, 2008, p. 939)

In other words, if the brain is constrained by computational tractability, then it limits the possible set of models which could be used when modeling cognition. Van Rooij specifically traces this back to the idea that the Church-Turing thesis is not merely a limitation of finitary computation, but is a limitation of reality as a whole, or, as van Rooij puts it, "The Church-Turing Thesis is a hypothesis about the state of the world" (van Rooij, 2008, p. 943).

Wolfram similarly applies his ideas specifically to the brain, saying:

> So what about computations that we perform abstractly with computers or in our brains? Can these perhaps be more sophisticated? Presumably they cannot, at least if we want actual results, and not just generalities. For if a computation is to be carried out explicitly, then it must ultimately be implemented as a physical process, and must therefore be subject to the same limitations as any such process (Wolfram, 2002, p. 721).

Thus, physicalism, when defined sufficiently to distinguish it from anything else, has been defined by its supporters as being equivalent to computationalism. This allows a more methodical examination of physicalism and dualism to determine which is likely to be true.

3   The Halting Problem

One of the classic unsolvable problems in computability is the "halting problem." In universal computation systems, there are ways to cause computations to repeat themselves. However, this leads to a possible problem—if a function is poorly written, the function may get caught in a repetitive portion and not be able to leave. This computation would be a non-halter, and therefore, left to itself, would never complete. Most familiar computations are halting computations, as demonstrated in the following computer program. All programming examples are given in JavaScript for readability.

```
function double(x) {
  var y;
  y = x * 2;
  return y;
}
```

Figure 5.2: A function to double a value

This program defines a function called *double* which obviously doubles its input. It creates a temporary variable called y to hold the result of the computation and then returns y as the final value for the function. So, after defining it, the function can be used by saying *double*(4) which would give 8, or *double*(z) which would take the value currently denoted by z and return whatever is double of z.

The next example will demonstrate the operation of a loop. This program computes the factorial of a number which is the result of multiplying a number by all of the numbers below it down to 1. For instance, $factorial(5)$ is $5 * 4 * 3 * 2 * 1$. $factorial(3)$ is $3 * 2 * 1$. So, the number of computations performed, while always finite for a finite number, varies with the value given. A typical way to program a factorial function follows:

```
function factorial(x) {
  var val;
  var multiplier;

  val = 1;
  multiplier = x;

  while(multiplier > 1) {
    val = val * multiplier;
    multiplier = multiplier - 1;
  }

  return val;
}
```

Figure 5.3: A function to compute the factorial of a number

This function defines two temporary variables—*val*, which holds the present state of the computation, and *multiplier*, which holds the next number that needs to be multiplied. Unlike algebraic systems, in most computer programming languages, variables do not have static values but can change over the course of the program. The $=$ is not an algebraic relationship, but rather it means assignment (e.g., $val = 1$ means that the value 1 is being assigned to the variable *val*).

In this program, the value of *multiplier* is set to the number given. Then the computation enters the loop. The *while* command tells the computer that while the value in the *multiplier* variable is greater than 1, it should perform the given computation contained in the curly braces. For example, if the function is performed with the value of 3, *multiplier* will be assigned the value 3, which is greater than 1. Then the computation within the *while* loop will be performed—it will multiply *val* (which starts off at 1) with *multiplier* (which is currently 3), and then assign that result back into *val*. *val* now has the number 3. *multiplier* is then decreased by one, and now has the value 2. The bracket indicates the end of the loop computation, so the condition is re-evaluated. *multiplier*'s value of 2 is still greater than one, so we perform the loop again. *val* (which is now 3) is multiplied by *multiplier* (which is now 2) and the value (6) is assigned back into *val*. *multiplier* is again decreased and is now 1. Now that the computation is at the end of the loop, the condition will be evaluated again, and this time *multiplier* is no longer greater than 1. Because the condition is no longer true, the loop does not run again, and the computation process goes on to the next statement.

The next statement returns the value in *val* as the result of the entire computation. Thus, since *val* currently holds 6, this function returns 6 as the result of *factorial*(3), which is the correct result. Since it does eventually return a value, it is considered a halting program. It will take longer to return a value if the input is bigger (since it has to run the loop computation process more times), and it will return invalid values if the input is less than one (or not an integer), but it will always return a value. Therefore, since it will always complete in a finite number of steps, it is a halter.

If the programmer writing this function forgot a step (e.g., to write the instruction that decreases *multiplier*), then instead of the previous program, the program might read as follows:

```
function factorial(x) {
  var val;
  var multiplier;

  val = 1;
  multiplier = x;
  while(multiplier > 1) {
    val = val * multiplier;
  }

  return val;
}
```

Figure 5.4: An incorrect function to compute the factorial of a number

In this example, since *multiplier* is never decreased, then, for any input greater than 1, this function will never stop computing! Therefore, in terms of the halting problem, it doesn't halt.

Functions on universal computation systems are convertible to numbers. In fact, that's how computers work—the computer stores the program as a very large number. One example of how this can work is that each character in the above program can be converted to a fixed-size number and then joined together to a large number to denote the program. And this is, in fact, how some programming languages function. Most of the time, however, the conversion of a program into a number actually works by doing a more intensive conversion of the program into a numeric language that the computer understands.

Nonetheless, in each case, the program gets converted into a (usually very large) number. Therefore, since any program can be converted into a counting number, there are only a countably infinite number of possible programs. But more importantly, it means that this program, since it is (or can be represented by) a number, can itself be an input to a function!

Some functions halt on certain inputs, but do not halt on other inputs. The halting question can only be asked on a combination of both the program and the input since some inputs may halt, and others may not. Therefore, the halting problem is a function of two variables—the program p and the input i. Every program/input combination either will halt, or it will not. There is no in-between state possible on finitary computations. Therefore, $H(p, i)$ can be denoted as a function which takes the program p and input i and gives as a result a 1 if $p(i)$ halts, or a 0 if $p(i)$ does not halt. This is known as a "decision problem"—a problem which takes inputs and decides if the inputs have a particular feature or match a given pattern. Interestingly, the program $H(p, i)$ cannot be formed using a universal computation system. This can be proved similarly to the early proof of incomputability.

To test this, first it must be assumed that $H(p, i)$ is a program that can be implemented with a universal computation system. If $H(p, i)$ can be implemented, then it can also be used by a longer program. A program which does this, $N(p)$, is described below:

```
function N(p) {
    if(H(p, p) == 1) {
        while(1 == 1) {
        }
    }

    return 1;
}
```

Figure 5.5: A theoretical function using the halting function which demonstrates its impossibility

This function starts by evaluating the halting problem of its input, p, given itself as the value. If the halting problem of a program p with itself as the input says "Yes it halts" (i.e., it gives a value of 1), an infinite loop (i.e., a computation which does not halt) will be performed. If not, the computation should return a value of 1, completing the computation (i.e., the program will halt with that value). One can ask the question, does $N(N)$ halt? If it does, then this program will loop forever, but it can't, because it has already been determined that it does not halt! Hence, a contradiction. Likewise the reverse. If $N(N)$ does not halt, then $N(N)$ will halt. Therefore, $H(p, i)$ cannot be solved using a universal computation system.

This process may seem like an equivocation on the nature of the functions being described since all of the programs so far have a single input while $H(p, i)$ has two inputs. However, any number of inputs can be encoded onto a single input using delimiters. Therefore, specifying multiple inputs is just an easier way to write out the function than the required steps for encapsulating the inputs together into a single value.

4   Turing Oracles as Solutions for Incomputable Problems

Turing recognized that although the value of $H(p, i)$ was not computable, it was in fact a true function of its variables—that is, for every input set, it yielded a single output. Thus, the halting problem was a hard problem—it had a solution, but not one that was determinable through finitary computation. Some important questions arose from this. Might there be other problems which are harder? Might there be problems which require the solution to the halting problem to figure out? If so, how does one go about reasoning about the computational difficulty of an unsolvable problem? The answer is in Turing Oracles.

A Turing Oracle (hereafter oracle) is a black-box function (i.e., no implementation description is given) which solves an incomputable function and yields its answer in a single step. An oracle machine is a combination of a normal computational system which also has access to an oracle. If the oracle is well-defined in its abilities, it can be used to reason about the process even if the process as a whole is incomputable. An oracle machine, then, is a regular machine (i.e., a normal computable function) which is connected to an oracle (i.e., the function has access to an operation which is incomputable).

Alan Turing describes the oracle machine as follows:

> Let us suppose that we are supplied with some unspecified means of solving number theoretic problems; a kind of oracle as it were. We will not go any further into the nature of this oracle than to say that it cannot be a machine. With the help of the oracle we could form a new kind of machine (call them o-machines), having as one of its fundamental processes that of solving a given number theoretic problem. (Turing, 1939, §4)

Even though the values of functions based on oracle machines cannot be computed (since they are by definition incomputable), it is still possible to reason about which problems are reducible to oracles and which oracles they are reducible to. Posed another way, if a programmer had an oracle for a given problem, what other problems could be solved? For instance, there is an incomputable function called Rado's Sigma

Function (affectionately known as the "busy beaver" function). This function says, given n, what is the longest non-infinite output of any program of size n? This is an incomputable function, but it can be shown to be computable given an oracle for $H(p, i)$.

If dualism is true, then at least some aspects of human cognition are not computable. However, given the discussion above, even if human cognition is partially incomputable, cognition may be at least representable if oracles are included in the allowable set of operations. Several researchers have previously discussed the possibility that the human mind may be an oracle machine (i.e., Copeland, 1998). However, none of them have suggested including oracles as a standard part of cognitive modeling, or how one might apply oracles to cognitive modeling (Bartlett, 2010a,b). The goal of this paper is to present the concept of modeling cognition via oracle machines and its application to a model of human problem-solving on insight problems.

5 Partial Solutions to Incomputable Functions Using Additional Axioms

Incomputable functions are unpredictably sensitive to initial conditions. In other words, there is no way to computably predict ahead of time the difference in behavior of the function from the differences in changes to the initial conditions. If this were possible, they would by definition not be incomputable! However, partial solutions to these functions can be made by incorporating additional axioms.

An axiom is a truth that is pre-computational. In other words, it is a truth about computation rather than a result of computation. Chaitin has shown that additional axioms can be used to make partial solutions of incomputable functions (Chaitin, 1982). For instance, if God were to say that there are 30 programs less than size n that halt for a given programming language, then that fact could be used to determine exactly which of those programs were the ones that halt. This is not a complete solution, but rather a partial solution. Nonetheless, it is a solution larger than what was originally determinable without the additional axiom.

Now, most axioms do not come in this form, but instead state that programs that have a certain pattern of state changes will never halt. This would not generate an exclusive list, but the list of additional programs that would be known non-halters through this axiom may be infinitely large. Therefore, by adding axioms, one could potentially be adding infinite subsets of solutions to incomputable problems. Axiom addition is also by definition non-algorithmic, for if axioms could be added algorithmically, then the halting problem would be solvable. Since this is not the case, axiom addition is not an algorithmic endeavor.

Once an axiom is known, however, then the computation of halters and non-halters for which sufficient axioms are known becomes an algorithmic problem. Therefore, the discovery of new axioms converts subsets of problems from non-algorithmic to algorithmic forms.

6   Towards Defining a Turing Oracle for Modeling Human Problem-Solving on Insight Problems

The next step after investigating computability theory is to relate this theory to problems in cognitive science—namely problem-solving for insight problems. Cognitive science usually breaks problem-solving into two broad categories—analysis problems and insight problems. Analysis problem are problems which can be solved using a known algorithm or set of known heuristics and are usually characterized by the subject being aware of how close he is to solving the problem, the benefits of continuous effort, and the use of pre-existing ideas to solve the problem. Insight problems, on the other hand, are problems which require a reconceptualization of the process in order to solve them (Chronicle et al., 2004).

An example of a classic insight problem is the nine-dot problem. In short, the problem is to take a 3x3 square of dots, and draw four lines that connect every dot without picking up the pencil. In order to solve the puzzle, the person must realize that the solution is to extend one of the lines beyond the confines of the box, and make a "non-dot turn." This reconceptualization of the problem is rare, the subject cannot gauge his or her own progress, and continuous effort is usually not as helpful as taking breaks.

Insight problems like these have significant structural similarity with incomputable functions. Incomputable functions can be partially solved through adding axioms to the mix. Axioms function a bit like reconceptualizations—they allow the problem to be worked from a different angle using a different approach. Because axioms cannot be generated algorithmically, it is difficult to conclude how close the solution is. Likewise, because the person is not following an algorithm (which is impossible for generating an axiom), continuous effort along the same path is not likely to be helpful.

Research on the nine-dot problem has shown that training on certain ideas such as non-dot turns in similar problems produces an increased success rate in solving the problem (Kershaw & Ohlsson, 2001; Kershaw, 2004). This effectively mirrors the way axioms function in mathematical problem-solving—by increasing the number of axioms available to the subject, experimenters were able to greatly reduce the difficulty of the nine-dot problem for participants.

Because it is mathematically impossible for a person to take an algorithmic approach to the general halting problem, it cannot be classed as an analysis problem. Because of this and its many similarities with other insight problems, the halting problem should be classified as an insight problem. As such, the discoveries that are made for how humans solve the halting problem will help formulate more generally a theory of human insight.

7   Human Solutions to the Halting Problem

As mentioned previously, if humans are able to solve incomputable functions, then the physicalism hypothesis is false.[2] The halting problem makes a good test case for this idea because it is one of the most widely studied class of incomputable problems on both a theoretical and a practical level.

Software development provides the first piece of insight into the process. In software development, humans have to develop software programs on universal computation systems, and those programs must halt. If they do not, their programs will be broken. Therefore, they must solve problems on at least some subset of the halting problem in order to accomplish their tasks. In addition, the problems that they are given to solve are not of their own design, so it is not a selection bias. It is simply not true that programmers are only choosing the programs to solve based on their intrinsic abilities to solve them because someone else (usually someone without the computational background needed to know the difference) is assigning the programs. In addition, it is incorrect to assert that programmers are working around their inabilities to solve certain types of halting problems, because, while the programmer might add some extrinsic complexity to a program, the complexity of the problem itself has an intrinsic minimum complexity regarding a given programming language. Likewise, simply writing it in another language does not help, because there exists a finite-sized transformer from any language to any other language, so being able to solve it in one language is de facto evidence of being able to solve it in another.

One may then conclude from the experience of the process of programming that significant evidence exists that humans are able to at least solve a similar problem to the halting problem. However, there are some important caveats.

A minor caveat is that machines in real life do not exhibit true universal computation as defined in the abstract. Universal computation systems have an infinite memory and can run forever without breaking down. However, there are two primary reasons why this is relatively unimportant to this discussion. The first is that even with fixed-size memory, the halting problem is still practically intractable. That is, the reason why fixed-size memories allow the halting problem to be solved is that a programmer could do an exhaustive search of machine states to determine if the machine state contains cycles (i.e., two exactly equivalent machine states) before halting. If so, then the program will halt. However, calculating the result of the halting problem even using finite-sized memory would require either enormous amounts of time or memory, on the order of 2^n, where n is the number of bits in memory.[3] In

---

[2]Penrose and others have suggested that physical processes are non-computational. However, they do so without a rigorous definition of what counts as "physical." The goal is to make the definition of physical rigorous enough to be testable, and therefore have used computational tractability as the requirement. See section 12 for additional discussion.

[3]As an example, one could solve the halting problem on fixed-size memories using a counter (Gurari, 1989). Since the number of possible machine states is 2^n, then if machine states are counted,

addition, the reasoning usually given by programmers as to why something should not halt is more similar to a proof than to an exhaustive set of attempts. If humans are regularly solving the halting problem for a large number of programs, then it is not because they are being aided by fixed-size computer memories.

The main caveat is that there exist programs (even very short programs) for which humans have not solved the halting problem. Many open problems in number theory can be quite simply converted into a halting problem so that the answer to the problem can be solved by knowing whether or not a given computation will halt. If humans have immediate access to a halting problem oracle, why do these programs give such trouble?

As an example, a perfect number is a number which is equal to the sum of its divisors excluding itself. For instance, 6 is a perfect number because 1, 2, and 3 are all divisors, and they add up to 6. It is not known if there are any odd perfect numbers. A program could be written to search and find an odd perfect number, and halt if it finds one. Such a program can be fairly simply expressed as:

```
function odd_perfect_divisor_exists() {
  var i = 3;

  while(true) { // This means to loop forever unless terminated
                // within the loop
    var divisors = all_divisors_of(i);
    var divisor_sum = sum(divisors);
    if(divisor_sum == i) {
      return i; // i.e. Halt
    } else {
      i = i + 2; // Go to the next odd number
    }
  }
}
```

Figure 5.6: A function which returns upon finding an odd perfect number

Therefore, if the above program halts, then there is an odd perfect number. If it does not halt, then there is not one. However, no human currently knows the answer to this question. Therefore, whatever it is that humans are doing, it is not directly knowing the answer to the halting problem.

---

we could determine that it must be a non-halting program if the program performs more than 2^n computations. A faster way of checking for cycles can be implemented, but it would generally require 2^n amount of memory.

Will humans ever be able to solve this problem? If humans possessed the same limitations on computation as computers, then they would never be able to solve this (and many other) problems. However, math and science, as disciplines, assume that unknown problems with definite answers will eventually be knowable. Simply stated, the progress of science depends on the ability of humans to eventually solve such problems as these.

In other words, if this is a fundamental limitation of humans, then the search for more and more mathematical truths may be madness—they will never be known. This has led some theorists such as Gregory Chaitin to suppose that theorists should, in some cases, simply assume the truth or falsity of some claims as axioms, even in absence of proofs of their truth (Chaitin, 2006). This seems to be a dangerous road to travel. Chaitin uses the fact that different geometries can be made from different axioms about the nature of the world to justify the arbitrariness of choosing axioms. In the case of geometry, for instance, the two different answers to the question of whether parallel lines can intersect generates two different geometries. However, choosing axiomatic truths for geometry is different than finding axiomatic truths for solving incomputable problems such as the halting problem, because in the former the axiom is unconstrained within the system and in the latter it is constrained but unprovable within the system. If an axiom is unconstrained, then given the remaining axioms, a fully consistent system can be maintained with either choice of axiom. In other words, either axiom is equally consistent with the remaining axioms. If an axiom is constrained but unprovable, then the truthfulness of an axiom is dependent on the remaining axioms. In other words, one axiom is true and another is false given the remaining axioms. In the case of reasoning about the halting problem, programmers are dealing entirely with constrained but unprovable axioms. It might be a worthwhile endeavor to provisionally accept an axiom and see where it leads, but it is dangerous to include a provisionally accepted axiom on equal ground with other types of axioms in formal mathematics.

Another option, however, is that humans are able to incrementally arrive at solutions to halting problems. This would mean that humans have access to an oracle which is more powerful than finitary computational systems, but less powerful than a halting oracle.

8   An Oracle for Insight Problems

Selmer Bringsjord has argued for the mind being hyper-computational on the basis of his research into human ability to solve the halting problem. His group claims that they could always determine the halting problem for Turing machines of size n if they took into account the determination of the halting problem for Turing machines of size $n - 1$ (Bringsjord, Kellet, Shilliday, & Taylor, 2006).

Bringsjord's group has considerable experience with the halting problem, but it is impossible to tell if his formulation is completely true based on the size of the problem space when n goes beyond 4—there are then too many programs for humans to analyze (when n is 5, there are 63,403,380,965,376 programs). What he found, though, is that his group could formulate halting proofs for programs of size n based on previous patterns which were identified for size $n - 1$. They used the proofs that they made for size $n - 1$ as a basis for the proofs in programs of size n. This is itself an interesting result, though it is hard to say that these are necessarily based on program size, since there is nothing in the halting problem that is program-size dependent. A better interpretation is that the proofs were built by introducing constrained axioms. The larger programs utilized the axioms introduced in smaller programs, but potentially required more axioms to solve. Therefore, the proofs utilized the smaller programs because they utilized the axioms demonstrated there. As the programs became larger, the number of axioms required to determine a solution also grew.

This explanation actually fits surprisingly well—it is non-algorithmic (it is determining unprovable axioms), it is incremental (each axiom gives more explanatory power), and it is weaker than a halting oracle.

To put this more formally, let's define some values:

A—the minimum set of axioms required to solve $Q(p, i)$

Q—a decision problem (such as the halting problem)

p—a program

i—the input to program p

B—a set of axioms such that the size of the set of the intersection of A and B is one smaller than A. In other words, B contains all of the axioms required to solve $Q(p, i)$ except one.

From these definitions human insight can be described by the following oracle:

$$A = I(Q, p, i, B) \tag{5.1}$$

In other words, if a human is given a decision problem over a certain input, and he or she knows all of the axioms needed to solve the problem except one, then human insight will reveal the remaining axiom. If true, this would explain why insight is both incremental and non-computational. It goes beyond what is available to computation, but still has prerequisites. In this proposal, all axioms are known except one. Thus, in the case of finding odd perfect numbers, the problem of finding the solution to the problem is that there are not enough pre-existing axioms to infer the final axiom.

9   Problems and Directions

The main problem with the description as stated is that there are different kinds of axioms, yet there is insufficient mathematical theory (at least known to the author) to differentiate types of axioms. At present, a distinction should be made between bottom-up and top-down axioms. As mentioned earlier, if God would say that there are x halting programs of size n, a programmer could determine which ones they were by running all of them simultaneously until x of them halt. This kind of axiom, a "top-down" axiom, requires prior knowledge of the entire spectrum of the problem to determine. Another kind of axiom, a "bottom-up" axiom, requires a minimum of understanding in order to be apprehended. Its truth is knowable even if not provable within its own formalism, and its application is not intrinsically bounded.

An example of a bottom-up axiom is an axiom which says that if a program has a loop whose control variable is monotonically decreasing and has a termination condition which is greater than its start value, then that program will never halt. That axiom, which is provable by induction, will then allow a programmer to determine the value of the halting problem for an infinite subset of programs.[4] Thus, it acts as a bottom-up axiom. In addition, as should be obvious, the introduction of such an axiom converts an infinite subset of problems from insight problems to analysis problems. Knowing such axioms allows the programmer to proceed algorithmically!

As a result, several open questions emerge:

1. Are there other properties of axioms which are important to the sequence in which they may be found?

2. Are there other prerequisites for finding these axioms?

3. In what ways (if any) do axioms relate to program size?

4. Is there a proper way to measure the size of an axiom?

Chaitin's proposal for measuring axioms is related to his Ω probability. Ω is the probability for a given Turing machine as to whether or not it will halt, which, for the sake of his theory, is written out as a string of bits. Chaitin measures axioms by the number of bits of Ω they are able to compute. If an axiom can deduce two bits of Ω, then the axiom is two bits long (Chaitin, 2007). A naive approach to using this definition might say that humans are able to deduce a single bit of Ω when needed. However, these bits are much "smaller" than the types of axioms that humans tend to develop, which are much wider in extent, as each bit of Ω is a single program, rather

---

[4]Some may claim that, since it is proved using an inductive proof, this statement becomes a theorem rather than an axiom. However, it is only a theorem from second-order logic, since general induction requires second-order logic and can only be imported to first-order logic as an axiom (Enderton, 2012). Since the machine itself is a first-order logic machine (Turing, 1936), it is an axiom from the perspective of the first-order system.

than a collection of programs. There seems to be, based on experience, some intrinsic ordering on the discoverability of axioms present within Ω. An algorithm can discover 1s (halts) within omega, with an implicit ordering based on length of program and the program's running time. For instance, a program could be written which started program 1 at time 1, program 2 at time 2, etc. Each iteration would run one cycle of each current program and start one new program. Each program that halts gives one bit of omega. Therefore, by exhaustive search, solutions to Ω can be discovered one bit at a time. However, this does not match the way humans arrive at the solution, which is a more generalized axiomatic approach (covering multiple cases of programs at a time—not just a single instance like Ω). Likewise, such algorithms can never discover the 0s (non-halters). Therefore, although Ω is a powerful conceptualization of the solution to the halting problem, it is unlikely to be helpful in the types of axioms that humans appear to be discovering.

Another possible way to measure the size of an axiom is to measure the size of the recognizer function needed to recognize instances of the axiom. But again, it is unclear whether or not that would be the measurement which would give the proper ordering of axiom determination. It may be harder to implement a recognizer than it is to intuitively recognize an axiom.

Therefore, in order to proceed further, additional research is needed into the nature of axioms themselves and the different ways that they can be categorized and quantified in order to find a natural sizing and ordering for them.

Again, two questions emerge that relate to the embodiment of the oracle itself:

1. How reliable is the axiom-finding oracle?

2. What are individual differences in this oracle?

The answers to such questions will lead to more understanding about how the oracle interacts with the rest of the mind's systems.

10   Generalizing the Oracle Method

Although important, the main focus of this paper is not the specific oracle outlined above. The larger point is that if an operation in the mind is non-physical, this does not preclude it from being modeled. Specifically, oracles seem to work for modeling a wide variety of non-physical operations. There are probably other operations which will require other formalisms, but formalisms should not be avoided simply because the formalism is not physically computable.

So how does one make a general application of oracles to modeling the mind? First of all, it is important that the operation under consideration be well-defined in terms of what it is doing. It is not worthwhile to simply state "something occurs here"—such is not a well-specified description. In the example above, specific preconditions (a decision problem, a program, its input, and a set of existing axioms) and a specific postcondition (the needed axiom to solve the problem) have been postulated. William Dembski's concept of specification could be used to determine whether or not the given specification is too broad or if it is reasonably constraining. Dembski's measurement is basically a relationship of the potentially described target states to the specification length. If a specification does not sufficiently limit the target space of possibilities, it is not a useful specification (Dembski, 2005).

Second, specific reasons must exist in order to believe that the proposed process is incomputable. Since solving the halting problem is known to be incomputable and adding axioms is incomputable by definition (otherwise they would be theorems), then specific evidence indicates that the proposed process is incomputable.

The hard part then comes in testing the theory. Because the results are incomputable, and not even likely reducible to a probability distribution, testing it is more difficult. In the case of computable causes, a specific end-point prediction can be established by computation, and then the result can be validated against that computation. In this case, the result is not computable, and therefore validation is more difficult. Validation will often be based on the qualitative description of the process rather than a quantitative prediction. Parts of it may still be quantifiable, but only with difficulty. For instance, to test the example presented, a method of identifying and counting the number of axioms within a person's mind is needed in order to come up with a quantifiable prediction. However, since this is not possible, it can only be tested based on secondary quantifications. Thus, testability on proposed oracles becomes much more dialectic.

11 Applications

This method of using oracles for modeling human cognition has many applications to both psychology and engineering, as well as to the history of technology. For psychology, it introduces a new way of evaluating mental causes and a new formalism for modeling and testing them. In several known cases, human problem solving outperforms what is expected from computationalism. For example, one group of researchers reported that human performance on the Traveling Salesman Problem scales linearly with the number of nodes, which far surpasses any computational estimator for the problem (Dry, Lee, Vickers, & Hughes, 2006). Therefore, modeling human performance in terms of an oracle machine may allow more accurate predictions of performance.

For engineering, oracles can be used to better identify and measure complexity. If axioms become quantifiable, and the number of axioms required to solve problems becomes quantifiable, then this can transform the practice of complexity estimation. One such method to use these ideas to calculate software complexity is given in Bartlett (2014).

This idea can also be applied to software computer games. Many computer games are organized by "levels" so that each level is harder than the previous one. One could use axioms as a measure of hardness and construct the levels so that each one introduces a new axiom used to complete the game. This would allow a more rigorous approach to game level design at least in certain types of computer games.

A final way of using this idea is in understanding the history of technology, including science and mathematics. It has been a curious feature that many "leaps" in scientific or mathematical thought have been made simultaneously by multiple people. Isaac Newton and Gottfried Leibniz both independently invented calculus, Gregory Chaitin and Andrey Kolmogorov both independently invented algorithmic information theory, Elisha Gray and Alexander Graham Bell both filed a patent for the telephone on the same day, and the list goes on and on (Aboites, Boltyanskii, & Wilson, 2012).[5] This model, if correct, would validate the view of T. D. Stokes that "even though there is no algorithm of discovery, there are logical elements present in the process whereby a novel hypothesis is devised." (Stokes, 1986, p. 111). This model presents a non-algorithmic process and shows the logical elements which are within its prerequisites. Therefore, when ideas are widespread, multiple people will each be a single axiom away from discovery. Consequently, faced with the same problem, many different people will be able to realize the same missing axiom.

12   Final Considerations

There is good evidence human cognition goes beyond what has been traditionally considered as "physical," and a lack of physicality does not preclude cognitive modeling. "Physical" has been defined as "computable" in order to avoid the ambiguities of the term. This is important because someone might try to assert that humans have a separate soul and that it is simply physical. Without a solid definition of what is and is not physical, nothing prevents such a formulation.

Roger Penrose and Jack Copeland have both made a similar suggestion (Copeland, 1998; Hodges, 2000). Both have agreed that humans seem to be oracle machines, but in a purely physical sense. However, neither of them provided a sufficient definition of what counted as physical or non-physical to make a proper distinction. Nothing that either of them has said would contradict what is defended in this paper, though Penrose argues that there is even more to human consciousness than is representable through oracle machines—a position also not in contradiction to the claims defended here. For instance, it is hard to consider the act of true understanding as a process involving numbers at all, as John Searle's Chinese Room argument shows (Searle, 1980).

Another possible objection, then, is to say that the universe as a whole isn't physical. It could be possible, for instance, that even the fundamental laws of matter are only fully describable using oracles, and none of them at all are computable with

---

[5]Appendix A of Aboites et al. (2012) contains quite an impressive list of codiscoveries.

finitary methods, and therefore finitary methods can only be used to solve certain macro-systems which are the exception rather than the rule. However, even if true, that would not lead to the conclusion that physicalism is true and incomputable functions should be classified as physical along with computable ones. Instead it would reveal that the idealists such as Richard Conn Henry (2005), who believe that the physical is a mere epiphenomenon and the non-physical is what is really real, were the ones who were right all along. Douglas Robertson (1999) comments:

> The possibility that phenomena exist that cannot be modeled with mathematics may throw an interesting light on Weinberg's famous comment: "The more the universe seems comprehensible, the more it seems pointless." It might turn out that only that portion of the universe that happens to be comprehensible is also pointless.

In any case, while it is certainly an improbable proposition, it is a logical possibility that physicalism is not true even for physics!

While the present discussion focuses on models of *human* insight, that limitation is purely practical—there is no known way of detecting or measuring insight behavior on non-human creatures—and there is no philosophical, theoretical, or theological reason why such processes could not be occurring in other creatures at a much lower level. Nothing in this proposal limits itself either to modeling humans or even organisms. However, in humans it seems most obvious and evident that restricting reality to only computable functions is incorrect.

References

Aboites, V., Boltyanskii, V. G., & Wilson, M. (2012). A model for co-discovery in science based on the synchronization of gauss maps. *International Journal of Pure and Applied Mathematics*, 79(2), 357–373. Available from http://ijpam. eu/contents/2012-79-2/15/15.pdf

Bartlett, J. (2010a). *A critique of nonreductive physicalism*. Phillips Theological Seminary, Tulsa, OK. Unpublished Master's Integrative Paper.

Bartlett, J. (2010b). Developing an approach to non-physical cognitive causation in a creation perspective. *Occasional Papers of the BSG*, 17, 3.

Bartlett, J. (2014). Calculating software complexity using the halting problem. In J. Bartlett, D. Halsmer, & M. R. Hall (Eds.), *Engineering and the ultimate* (pp. 123–130). Broken Arrow, OK: Blyth Institute Press.

Bringsjord, S., Kellett, O., Shilliday, A., Taylor, J., van Heuveln, B., Yang, Y., Baumes, J., & Ross, K. (2006). A new Gödelian argument for hypercomputing minds based on the busy beaver problem. *Applied Mathematics and Computation*, 176(2), 516–530. Available from http://citeseerx.ist.psu.edu/ viewdoc/download?doi=10.1.1.91.5786&rep=rep1&type=pdf

Chaitin, G. (1982). Gödel's theorem and information. *International Journal of Theoretical Physics*, 21, 941–954. Available from http://www.cs.auckland.ac.nz/ ~chaitin/georgia.html

Chaitin, G. (2006). The limits of reason. *Scientific American*, 294(3), 74–81. Available from http://www.umcs.maine.edu/~chaitin/sciamer3.pdf

Chaitin, G. (2007). The halting problem omega: Irreducible complexity in pure mathematics. *Milan Journal of Mathematics*, 75(1), 291–304. Available from http://www.cs.auckland.ac.nz/~chaitin/mjm.html

Chronicle, E. P., MacGregor, J. N., & Ormerod, T. C. (2004). What makes an insight problem? The roles of heuristics, goal conception, and solution recoding in knowledge-lean problems. *Journal of Experimental Psychology: Learning, Memory, and Cognition*, 30(1), 14–27. Available from http://citeseerx.ist.psu. edu/viewdoc/download?doi=10.1.1.122.5917&rep=rep1&type=pdf

Church, A. (1936). An unsolvable problem of elementary number theory. *American Journal of Mathematics*, 58(2), 345–363. Available from http://www.jstor. org/stable/2371045

Copeland, B. J. (1998). Turing's o-machines, Searle, Penrose, and the brain. *Analysis*, 58(2), 128–138. Available from http://www.hums.canterbury.ac.nz/phil/people/personal_pages/jack_copeland/pub/turing1.pdf

Dembski, W. A. (2005). Specification: The pattern that signifies intelligence. *Philosophia Christi*, 7(2), 299–343. Available from http://www.designinference.com/documents/2005.06.Specification.pdf

Dry, M., Lee, M. D., Vickers, D., & Hughes, P. (2006). Human performance on visually presented traveling salesperson problems with varying numbers of nodes. *Journal of Problem Solving*, 1(1), 20–32. Available from http://docs.lib.purdue.edu/cgi/viewcontent.cgi?article=1004&context=jps

Enderton, H. B. (2012). Second-order and higher-order logic. In E. Zalta (Ed.), *The Stanford encyclopedia of philosophy*. The Metaphysics Research Lab, fall 2012 edition. Available from http://plato.stanford.edu/archives/fall2012/entries/logic-higher-order/

Gurari, E. (1989). *An introduction to the theory of computation*. New York: Computer Science Press. Available from http://www.cse.ohio-state.edu/~gurari/theory-bk/theory-bk-twose6.html

Heart, W. D. (1994). Dualism. In S. Guttenplan (Ed.), *A companion to the philosophy of mind* (pp. 265–269). Malden, MA: Wiley-Blackwell.

Henry, R. C. (2005). The mental universe. *Nature*, 436, 29.

Hodges, A. (2000). Uncomputability in the work of Alan Turing and Roger Penrose. Available from http://www.turing.org.uk/philosophy/lecture1.html

Horgan, T. (1994). Physicalism. In S. Guttenplan (Ed.), *A companion to the philosophy of mind* (pp. 471–479). Malden, MA: Wiley-Blackwell.

Kershaw, T. C. (2004). Key actions in insight problems: Further evidence for the importance of non-dot turns in the nine-dot problem. In *Proceedings of the twenty-sixth annual conference of the cognitive science society* (pp. 678–683). Available from http://www.cogsci.northwestern.edu/cogsci2004/papers/paper156.pdf

Kershaw, T. C. & Ohlsson, S. (2001). Training for insight: The case of the nine-dot problem. In *Proceedings of the twenty-third annual conference of the cognitive science society* (pp. 489–493). Available from http://conferences.inf.ed.ac.uk/cogsci2001/pdf-files/0489.pdf

O'Connor, T. & Wong, H. Y. (2012). Emergent properties. In N. Zalta (Ed.), *The Stanford encyclopedia of philosophy*. The Metaphysics Research Lab, spring 2012 edition. Available from http://plato.stanford.edu/archives/spr2012/entries/properties-emergent/

Robertson, D. S. (1999). Algorithmic information theory, free will, and the Turing test. *Complexity*, 4(3), 25–34. Available from http://cires.colorado.edu/~doug/philosophy/info8.pdf

Searle, J. R. (1980). Minds, brains, and programs. *The Behavioral and Brain Sciences*, 3(3), 417 457. Available from http://cogprints.org/7150/1/10.1.1.83.5248.pdf

Stokes, T. D. (1986). Reason in the zeitgeist. *History of Science*, 24, 111–123. Available from http://adsabs.harvard.edu/full/1986HisSc..24..111S

Stoljar, D. (2009). Physicalism. In E. Zalta (Ed.), *The Stanford encyclopedia of philsophy*. The Metaphysics Research Lab, fall 2009 edition. Available from http://plato.stanford.edu/archives/fall2009/entries/physicalism/

Turing, A. M. (1936). On computable numbers, with application to the entscheidungsproblem. *Proceedings of the London Mathematical Society*, 42, 230–265. Available from http://www.cs.virginia.edu/~robins/Turing_Paper_1936.pdf

Turing, A. M. (1937). Computability and λ-definability. *The Journal of Symbolic Logic*, 2(4), 153–163. Available from http://www.jstor.org/stable/2268280

Turing, A. M. (1939). Systems of logic based on ordinals. *Proceedings of the London Mathematical Society*, 45. Available from https://webspace.princeton.edu/users/jedwards/Turing%20Centennial%202012/Mudd%20Archive%20files/12285_AC100_Turing_1938.pdf

van Rooij, I. (2008). The tractable cognition thesis. *Cognitive Science: A Multidisciplinary Journal*, 32(6). Available from http://staff.science.uva.nl/~szymanik/papers/TractableCognition.pdf

Wolfram, S. (2002). *A new kind of science*. Champaign: Wolfram Media. Available from http://www.wolframscience.com/nksonline/

Calculating Software Complexity Using the Halting Problem

6 ||

JONATHAN BARTLETT

The Blyth Institute

Abstract

Calculating the complexity of software projects is important to software engineering as it helps in estimating the likely locations of bugs as well as the number of resources required to modify certain program areas. Cyclomatic complexity is one of the primary estimators of software complexity which operates by counted branch points in software code. However, cyclomatic complexity assumes that all branch points are equally complex. Some types of branch points require more creativity and foresight to understand and program correctly than others. Specifically, when knowledge of the behavior of a loop or recursion requires solving a problem similar to the halting problem, that loop has intrinsically more complexity than other types of loops or conditions. Halting-problem-like problems can be detected by looking for loops whose termination conditions are not intrinsically bound in the looping construct. These types of loops are counted to find the program complexity. This metric is orthogonal to cyclomatic complexity (which remains useful) rather than as a substitute for it.

1 Complexity Metrics in Software

Managing software development is often about managing risks - knowing which tasks are likely to take more time than others, which features are more likely to impact others, how much testing will be required to make sure that a feature is solid, and whether a bug fix or a feature implementation requested right before release will be more likely to make the code more stable or lead to other bugs.

One of the key considerations of risk management is software complexity. Complex software is inherently more difficult to build, test, and maintain. Therefore, it is critical for software development managers to know which parts of code are most complex and therefore more likely to incur failures if modified.

2   A Brief History of Software Complexity Metrics

Early complexity metrics were based almost entirely on the amount of code produced. Therefore, a function which contained 10 lines of code was considered more complex than one which contained only 5. However, since "lines" of code often varies due to stylistic differences, Halstead developed a set of measures based on the number of operators and operands present within the code (Kearney et al., 1986).

Lines of code and related metrics are still often used for software effort estimation, but its use in analyzing code complexity has fallen away. It quickly became clear that not all code is the same, and some operators are inherently more complex than others. Specifically, the decision structure of the program was inherently more complex than the computations. Cyclomatic complexity was created to measure the size of the decision structure of the program (McCabe, 1976). This is done by creating a graph of all basic code blocks as nodes and then adding edges which show how control can move between them. The formula for calculating the complexity is:

$$E - N + T \qquad (6.1)$$

E is the number of edges, N is the number of nodes, and T is the number of terminating nodes (entry and exit points - usually 2).

Cyclomatic complexity is extremely useful in determining how to test software. The cyclomatic complexity of a program is also the minimum number of tests needed to cover every control flow branch of a program. If the cyclomatic complexity of a program is 5, then at least 5 tests must be devised to test every branch of code. Such tests do not guarantee total coverage of all possible test conditions, but they will verify that every statement in the program will contribute to at least one test.

The ABC metric is a simplified metric combining aspects of both lines of code and cyclomatic complexity. It works by simply counting the number of assignment statements, the number of branches (direct flow control shifts - i.e., function calls), and the number of conditionals (Fitzpatrick, 1997, pp. 2–3). These can then be analyzed on a whole program, per-module, or per-function basis, to give an overview of complexity and size of a software program.

3   Deeper Difficulties in Software

While each of the previously-mentioned metrics have their usefulness, none of them get at the deeper difficulties that make software projects complex. Many programming languages attempt to remove the inherent difficulties within software. Some, like COBOL, attempt to remove the mathematical notation common to programming languages. Others, like Java, try to simplify software by encapsulating related methods into objects. Visual programming languages, such as Flowcode, turn all software into visual representations, allowing the user to drag and drop flowchart-like components to accomplish programming tasks. The idea is that it is the text-based nature of the software which causes complexity and confusion within software.

While each of these actually do relieve certain specific problems in software development, none of them are able to remove the complexity of software development because the complexity is inherent in the nature of software development itself. What makes software development difficult is its open-ended nature. Most general-purpose programming languages today are universal in nature - that is, they can perform any computable function that any other programming language can perform. Thus, they are open-ended - the types of operations that they perform are entirely specifiable by the programmer and are not restricted. Universal programming languages are chaotic - that is, there is no easy mapping between programs, data, and results. Therefore, predicting the output over a wide swath of code and data can be difficult.

Languages are chaotic because of arbitrary looping structures. Interestingly, this is precisely the same part that causes it to be open-ended. Arbitrary looping structures allow a programmer to generate any possible computable function. As such, they also allow a programmer to write programs whose results are chaotic. In practical terms, arbitrary looping structures are `while` statements, `jump/goto` statements, recursive functions, and continuations, though others may be possible.

Occasionally, the solution to this problem has been to reduce the scope of the language. SuperGlue is one language which is specifically designed to be as expressive as possible while avoiding constructs that lead to complexity (McDirmid, 2006). However, ultimately, to get beyond the originally conceived computational bounds of the programming language, universality, as well as the complexity that goes with it, are required.

4   The Halting Problem as an Insight Problem

As discussed elsewhere in this volume, some problems are not amenable to analytical analysis and require insight in order to solve them (Bartlett, 2014; Holloway, 2014). Neither are such problems computable - no computer is capable of calculating these problems. One such problem that is relevant is the halting problem.

The halting problem states that, given a universal language, there is no program that can be written which will tell if any arbitrary program written in that language will ever complete (i.e., halt). This is not based on the size of the program code, but rather on the nature of the constructs available. This is not to say that one could not write a program to tell if certain subsets of programs written in that language will halt, but there could not be a program to tell if any given program would halt.

What is intriguing about this, as noted by Bartlett (2014), is that computer programmers seem to be able to possess this power to some degree. Since the complexity of problems assigned to them are arbitrarily hard (i.e., management, not the programmer, often decides what must be done), and the reason that arbitrarily hard programs are possible is because of the open-ended nature of universal programming languages (the parts of the language which *create* the chaos are precisely the ones *required* for programs of arbitrary complexity), it can be said that human programmers are generally reliable halting problem solvers.

There are cases (usually coming from number theory) where it is not known whether or not programs (even very simple ones!) halt. These seem to indicate that there are different levels of difficulty and different levels of insight required to make determinations. Some might even use these cases to argue against the general ability of humans to reliably solve the halting problem. However, the advancement of science and mathematics actually depends on the ability of humans to be able to accomplish such tasks. In other words, if the ability of humans to figure out such problems is doubted, then the progress of science itself is brought into question. Should mathematicians stop looking for answers in number theory? Or is it better to assume that the proper insight will come one day? The ability of programmers to reliably solve halting problems in their daily work should lend hope to the mathematicians that someone will eventually be able to have the insight to solve their problems as well.

5 Using the Halting Problem to Measure Software Complexity

The trouble with insight problems is that they are not reliably solved by individuals. They are unreliably solvable - in other words, a solution is possible, but there is no analytic procedure to do so. Even worse, if programmers do not realize that they are looking at an insight problem, they might not know that special care must be taken.

So how is an insight problem in code recognized? Since the types of programming structures which make a program universal have already been determined (i.e., arbitrary looping structures), those structures in code can therefore be detected.

For most programming languages, the main structures which must be detected are the following:

1. Loops where the iterations are not implicit in the control structure (called *open-ended loops*)

2. Recursive functions (which are just another way of implementing open-ended loops)

For open-ended loops, consider the following two programs that each print out the square of every number in an array:

```
ary.each{|x|
  puts x * x
}
```

Figure 6.1: A program with an implicitly-terminating loop

```
i = 0
while(i < a.length) {
  puts x * x
  i = i + 1
}
```

Figure 6.2: A program with an arbitrary looping structure

The implementation in Figure 6.2 is more complex than the one in Figure 6.1, but not because of the size of the program. What makes it more complex is that it utilizes an arbitrary looping structure - the while statement. In Figure 6.1, the looping is inherently bound by the loop operator. In Figure 6.2, the programmer must specifically act to make the loop terminate appropriately. There is no way an each statement on its own will fail to halt. There are many ways in which a while statement can fail to halt. The termination is decoupled from the loop construct itself.

Therefore, as a first pass to measure software complexity, the programmer can simply count the number of open-ended looping structures (either as loops or recursive functions) which occur in the program.

6   Adding Axioms to Minimize Insight

However, as is obvious from Figure 6.2, there are many well-understood conventions which mitigate the complexity of certain kinds of open-ended looping structures. In that figure, for instance, the variable i starts at zero, monotonically increases, and then terminates at a predetermined stopping point, which will result in the loop's termination. Even in cases where this sequence of steps is not codified within the language using a special statement or procedure, it is a well-understood looping convention. If the convention is followed correctly, the loop will terminate.

Gregory Chaitin formalized this idea in his algorithmic information theory. He pointed out that while certain problems were unsolvable given a base set of axioms, by incorporating additional axioms into the problem, solutions can be found. For instance, following Chaitin, if God were to tell us how many programs of size N halted, that information could be used to solve the halting problem for programs of size N (Chaitin, 1982).[1]

In the same way, when a convention for constraining repetition is discovered, it can be incorporated into a canon of axioms which are also known to halt. And thus it should be treated almost on the same level as a language construct which produces close-ended loops, because language constructs enforce the validity of the axiom structurally, while conventions require the programmer to manually follow the convention correctly.

This canon of axioms can be codified into an extensible static analysis tool to check program complexity. Such a tool could consist of the set of potential non-terminating constructs, as well as a "book of conventions" which are the known conventions for ensuring termination. The tool would then measure the potential number of non-terminating constructs which do not conform to a pattern in the "book of conventions."[2]

In addition to the constructs which can be statically analyzed, some conventions (often termed as "patterns") will not be easily amenable to inference by software. They should, however, at least be documented, and they can be manually marked or removed after the fact. However, if the construct is not amenable to static analysis, extra effort should be taken to review all implementations of the construct manually.

It should also be recognized that these axioms should be treated as first-class insights—that is, the "book of conventions" should be considered a set of valuable intellectual assets. As solutions to insight problems, the "book of conventions" is by definition a set of solutions which are not immediately obvious, and, therefore, if a convention is not recorded, and is therefore "lost," it could well be a permanent loss of insight for an organization.

---

[1] For an informal proof of this, consider that the issue that makes the halting problem difficult is that if a program is running indefinitely, it is impossible to tell whether or not it is just taking a long time and will finish eventually, or if it truly will never finish. However, if it is known that k programs of size N will halt, one can simply run all programs of size N simultaneously. As long as the number of programs that have finished is smaller than k, then there are some programs in that set which will halt. Once all k programs finish, then it is needless to continue to run the remaining programs since it was given that exactly k programs finish. Therefore, if the number of programs in a set which halt is known ahead of time, it is possible to determine the answer to all the halting problems in a finite length of time—the length of time will be the maximum runtime of the longest running halting program.

[2] A similar procedure was independently developed by Bringsjord et al. (2006), Hertel (2009), and Harland (2007), though differing in many aspects and applications. Their solutions were to categorize non-halters rather than halters, and to do it based on runtime patterns rather than a static analysis of structural patterns in the program. They identified well-known patterns, data-mined for others, and then used a symbolic induction prover to match potential programs with these patterns. In addition, their purpose was for answering questions about computer science theory (specifically, the busy beaver problem) rather than assessing program complexity.

7   Using the Metric

The ultimate goal of the metric is to reduce the complexity of the software to zero. When a pattern which solves a restricted subset of the halting problem is discovered, it can be incorporated as a new axiom into the "book of conventions." Therefore, if there are any areas in the program which are marked as being complex, that means that it is still not known if the program will even finish! If a developer has a new insight into why a certain section of code will finish, this should be documented in the "book of conventions." If programmers cannot state why they think that the program will finish, it should be reviewed or rewritten. If the code cannot be reworked and the program cannot be proven to terminate, it should be considered highly suspicious.

In addition, areas of code which are complex given the constructs, but found in the "book of conventions" should be flagged for a second-pass review to make sure the conventions were followed appropriately. Such sections should also be flagged for programmers making modifications to be sure that their modifications do not upset the assumptions of the conventions.

8   Further Considerations

While this metric is very useful, it is obviously not the last word on complexity metrics. It does not technically supersede the other complexity metrics mentioned. Counting and estimating lines of code are still useful planning tools. Cyclomatic complexity is still a useful test coverage tool. However, this metric can be useful in identifying programming patterns which are intrinsically problematic and help mitigate possible problems with documentation, code review, and testing.

For future development, similar ideas could be applied not just to the halting complexity, but also to the complexity that variables are derived from. When the value of a variable is determined by multiple loops, or conditions within loops, or other sorts of non-linear mechanisms, the value of variables can be chaotic, even when they are not themselves what determines if the problem halts. Extending these ideas to variable calculation could allow for an even more comprehensive look at where program complexity lies.

References

Bartlett, J. (2014). Using Turing oracles in cognitive models of problem-solving. In J. Bartlett, D. Halsmer, & M. R. Hall (Eds.), *Engineering and the ultimate* (pp. 99–122). Broken Arrow, OK: Blyth Institute Press.

Bringsjord, S., Kellett, O., Shilliday, A., Taylor, J., van Heuveln, B., Yang, Y., Baumes, J., & Ross, K. (2006). A new Gödelian argument for hypercomputing minds based on the busy beaver problem. *Applied Mathematics and Computations*, 176(2), 516–530. Available from http://citeseerx.ist.psu.edu/viewdoc/download?doi=10.1.1.91.5786&rep=rep1&type=pdf

Chaitin, G. (1982). Gödel's theorem and information. *International Journal of Theoretical Physics*, 21, 941–954. Available from http://www.cs.auckland.ac.nz/~chaitin/georgia.html

Fitzpatrick, J. (1997). Applying the ABC metric to C, C++, and Java. *C++ Report*. Available from http://www.softwarerenovation.com/ABCMetric.pdf

Harland, J. (2007). Analysis of busy beaver machines with inductive proofs. In J. Gudmundsson & B. Jay (Eds.), *Cats'07: Proceedings of the 13th Australasian symposium on theory of computing* (pp. 71–78).

Hertel, J. (2009). Computing the uncomputable rado sigma function: An automated, symbolic induction prover for nonhalting Turing machines. *The Mathematica Journal*, 11(2), 270–283. Available from http://www.mathematica-journal.com/issue/v11i2/contents/Hertel/Hertel.pdf

Holloway, E. (2014). Complex specified information (CSI) collecting. In J. Bartlett, D. Halsmer, & M. R. Hall (Eds.), *Engineering and the ultimate* (pp. 153–166). Broken Arrow, OK: Blyth Institute Press.

Kearney, J. K., Sedlmeyer, R. L., Thompson, W. B., Gray, M. A., & Adler, M. A. (1986). Software complexity measurement. *Communications of the ACM*, 29(11), 1044–1050. Available from http://sunnyday.mit.edu/16.355/kearney.pdf

McCabe, T. J. (1976). A complexity measure. In *Proceedings of the 2nd international conference on software engineering*. Available from http://www.literateprogramming.com/mccabe.pdf

McDirmid, S. (2006). Turing completeness considered harmful: Component programming with a simple language. Submitted for publication. Available from http://lampwww.epfl.ch/~mcdirmid/papers/mcdirmid06turing.pdf

7

Algorithmic Specified Complexity

WINSTON EWERT, WILLIAM A. DEMBSKI, AND
ROBERT J. MARKS II

Baylor University
Discovery Institute

Abstract

Engineers like to think that they produce something different from that of a chaotic system. The Eiffel tower is fundamentally different from the same components lying in a heap on the ground. Mt. Rushmore is fundamentally different from a random mountainside. But engineers lack a good method for quantifying this idea. This has led some to reject the idea that engineered or designed systems can be detected. Various methods have been proposed, each of which has various faults. Some have trouble distinguishing noise from data, some are subjective, etc. For this study, conditional Kolmogorov complexity is used to measure the degree of specification of an object. The Kolmogorov complexity of an object is the length of the shortest computer program required to describe that object. Conditional Kolmogorov complexity is Kolmogorov complexity with access to a context. The program can extract information from the context in a variety of ways allowing more compression. The more compressible an object is, the greater the evidence that the object is specified. Random noise is incompressible, and so compression indicates that the object is not simply random noise. This model is intended to launch further dialog on use of conditional Kolmogorov complexity in the measurement of specified complexity.

1 Introduction

Intuitively, humans identify objects such as the carved faces at Mount Rushmore as qualitatively different from that of a random mountainside. However, quantifying this

131

Figure 7.1: The faces of Mount Rushmore—Public Domain

concept in an objective manner has proved difficult. Both mountainsides are made up of the same material components. They are both subject to the same physical forces and will react the same to almost all physical tests. Yet, there does appear to be something quite different about Mount Rushmore. There is a special something about carved faces that separates it from the rock it is carved in.

This "special something" is information. Information is what distinguishes an empty hard disk from a full one. Information is the difference between random scribbling and carefully printed prose. Information is the difference between car parts strewn over a lawn and a working truck.

While humans operate using an intuitive concept of information, attempts to develop a theory of information have thus far fallen short of the intuitive concept. Claude Shannon developed what its today known as Shannon information theory (Shannon et al., 1950). Shannon's concern was studying the problem of communication, that of sending information from one point to another. However, Shannon explicitly avoided the question of the meaningfulness of the information being transmitted, thus not quite capturing the concept of information as defined in this paper. In fact, under Shannon's model a random signal has the highest amount of information, the precise opposite of the intuitive concept.

Another model of information is that of algorithmic information theory (Chaitin, 1966; Solomonoff, 1960; Kolmogorov, 1968b). Techniques such as Kolmogorov complexity measure the complexity of an object as the minimum length computer program required to recreate the object; Chaitin refers to such minimum length programs as *elegant* (Chaitin, 2002). As with Shannon information, random noise is the most complex because it requires a long computer program to describe.

In contrast, simple patterns are not complex because a short computer program can describe the pattern. But neither simple patterns nor random noise are considered conceptual information. As with Shannon information, there is a disconnect between Kolmogorov complexity and conceptual information.

Other models are based on algorithmic information theory, but also take into account the computational resources required for the programs being run. Levin complexity adds the log of the execution time to the complexity of the problem (Levin, 1976). *Logical depth*, on the other hand, is concerned with the execution time of the shortest program (Bennett, 1988). There is a class of objects which are easy to describe but expensive to actually produce. It is argued (Bennett, 1988) that objects in this class must have been produced over a long history. Such objects are interesting, but do not seem to capture the intuitive concept of information in its entirety. English text or Mount Rushmore correspond to what is usually considered as information, but it is not clear that they can be most efficiently described as long running programs.

One approach to information is *specified complexity* as expressed by Dembski (Dembski, 1998). Dembski's concern is that of detecting design, the separation of that which can be explained by chance or necessity from that which is the product of intelligence. In order to infer design, an object must be both complex and specified. Complexity refers, essentially, to improbability. The probability of any given object depends on the chance hypothesis proposed to explain it. Improbability is a necessary but not sufficient condition for rejecting a chance hypothesis. Events which have a high probability under a given chance hypothesis do not give a reason to reject that hypothesis.

Specification is defined as conforming to an independently given pattern. The requirement for the pattern to be independent of the object being investigated is fundamental. Given absolute freedom of pattern selection, any object can be made to seem specified by selecting that object as the pattern. It is not impressive to hit a bullseye if the bullseye is painted on after the arrow has hit the wall. It is impressive to hit the bullseye if the bullseye was painted before the arrow was fired.

Investigators are often not able to choose the target prior to investigating the object. For example, life is a self-replicating process, and it would seem that an appropriate specification would be self-replication. Self-replication is what makes life such a fascinating area of investigation as compared to rocks. Human beings know about self-replication *because of* their knowledge of life, not as an independent fact. Therefore, it does not qualify as an independent specification.

The same is true of almost any specification in biology. It is tempting to consider flight a specification, but the pattern of flight would only be defined because flying animals have been observed. As with life in general, specific features in biology cannot be specified independently of the objects themselves.

The concept of specification has been criticized for being imprecisely defined and unquantifiable. It has also been charged that maintaining the independence of the patterns is difficult. But specification has been defined in a mathematically rigorous manner in several different ways (Dembski, 1998, 2002, 2005). Kolmogorov complexity, or a similar concept, is a persistent method used in these definitions. The goal of this paper is to present and defend a simple measure of specification that clearly alleviates these concerns. Towards this end, the authors propose to use *conditional Kolmogorov complexity* to quantify the degree of specification in an object. Conditional Kolmogorov complexity can then be combined with complexity as a measurement of specified complexity. This approach to measuring specified complexity is called *algorithmic specified complexity.*

As noted, Kolmogorov complexity has been suggested as a method for measuring specification. The novelty in the method presented here is the use of conditional Kolmogorov complexity. However, this paper also elucidates a number of examples of algorithmic compressibility demonstrating wider applicability than is often realized.

2   Method

2.1   Kolmogorov

Kolmogorov complexity is a method of measuring information. It is defined as the minimum length computer program, in bits, required to produce a binary string.

$$K(X) = \min_{U(p,)=X|p\in P} |p| \tag{7.1}$$

where

- $K(X)$ is the Kolmogorov complexity of X

- P is the set of all possible computer programs

- $U(p,)$ is the output of program p run without input

The definition is given for producing binary strings.

Kolmogorov complexity measures the degree to which a given bitstring follows a pattern. The more a bitstring follows a pattern, the shorter the program required to reproduce it. In contrast, if a bitstring exhibits no patterns, it is simply random, and a much longer program will be required to produce it.

Consider the example of a random binary string, 100100000010100000001010. It can be produced by the following Python program:

```
print '100100000010100000001010'
```

Figure 7.2: A Python program to produce an unpatterned bitstring

In contrast, the string 000000000000000000000000 can be produced by

```
print '0' * 24
```

Figure 7.3: A Python program to produce a patterned bit-string

Both strings are of the same length, but the string following a pattern requires a shorter program to produce; thus, a technique exists for measuring the degree to which a binary string follows a pattern.

Specification is defined as following an independently given pattern. Kolmogorov complexity provides the ability to precisely define and quantify the degree to which a binary string follows a pattern. Therefore, it seems plausible that a specification can be measured using Kolmogorov complexity. The more compressible a bitstring, the more specified it is.

However, Kolmogorov complexity seems unable to capture the entirety of what is intended by specification. Natural language text is not reducible to a simple pattern; however, it is an example of specification. The design of an electronic circuit should also be specified, but it is not reducible to a simple pattern. In fact, the cases of specification that Kolmogorov complexity seems able to capture are limited to objects which exhibit some very simple pattern. But these are not the objects of most interest in terms of specification.

There is also an extension of Kolmogorov complexity known as *conditional Kolmogorov complexity* which can be used (Kolmogorov, 1968a). With conditional Kolmogorov complexity, the program now has access to additional data as its input.

$$K(X|Y) = \min_{U(p,Y)=X|p\in P} |p| \tag{7.2}$$

where $U(p, Y)$ is the output of running program p with input Y.

In this calculation, the input provides additional data to the program. As a result, the program is no longer restricted to exploiting patterns in the desired output but can take advantage of the information provided by the input. Henceforth, this input is referred to as the *context*.

The use of context allows the measure to capture a broader range of specifications. It is possible to describe many bitstrings by combining a short program along with the contextual information. A useful range of specifications can be captured using this technique.

2.2 Algorithmic Specified Complexity

The following formula for algorithmic specified complexity (ASC) combines the measurement of specification and complexity.

$$A(X, C, p) = -\log p(X) - K(X|C) \tag{7.3}$$

where

- X is the bitstring being investigated

- C is the context as a bitstring

- p is the probability distribution which it is supposed that X has been selected from

- $p(X)$ is the probability of X occurring according to the chance hypothesis under consideration

Since high compressibility corresponds to specification, the compressed length of the string is subtracted. Thus, high improbability counts for specified complexity, but incompressible strings count against it.

For this number to become large requires X to be both complex (i.e., improbable) and specified (i.e., compressible). Failing on either of these counts will produce a low or negative value. Since Kolmogorov complexity can, at best, be upper bounded, the ASC can, at best, be lower bounded.

At best this measure can reject a given probability distribution. It makes no attempt to rule out chance-based hypotheses in general. However, it can conclude that a given probability distribution does a poor job in explaining a particular item. The value of ASC gives a measure of the confidence available for rejecting a chance hypothesis.

2.3   Functionality

Perhaps the most interesting form of specification is that of functionality. It is clear that machines, biological structures, and buildings all have functionality, but quantifying that functionality in an objective manner has proven difficult. However, ASC provides the ability to do this.

Any machine can be described in part by tests that it will pass: The functionality of a car can be tested by seeing whether it accelerates when the gas or brake pedals are pushed; the functionality of a cell by seeing whether it self-replicates. A test, or a number of tests, can be defined to identify the functionality of an object. The existence of a test supplies the ability to compress the object. Consider the following pseudocode program.

```
counter = 0
for each possible building design
    if building won't fall over
        counter += 1
        if counter == X
            return building design
```

Figure 7.4: A pseudocode program which uses a functional test to compress the specification of an object by its functionality

This program will output the design for a specific building based on a given value for X. Different values of X will produce different buildings. But any building that will not fall over can be expressed by this program. It may take a considerable amount of space to encode this number. However, if few designs are stable, the number will take much less space than what would be required to actually specify the building plans. Thus, the stability of the building plan enables compression, which in turn indicates specification.

Kolmogorov complexity is not limited to exploiting what humans perceive as simple patterns. It can also capture other aspects such as functionality. Functionality can be described as passing a test. As a result, functional objects are compressible.

3 Examples

3.1 Natural Language

Consider the sentence: "The quick brown fox jumps over the lazy dog." This sentence can be encoded as UTF-32, a system for encoding that allows the encoding of symbols from almost any alphabet. Since each character takes 32 bits, the message will be encoded as a total of 1,376 bits. In this example, the context will be taken to be the English alphabet along with a space. This is a minimal level of information about the English language.

To specify one of the 27 characters requires $\log_2 27$ bits. To specify the 43 characters in the sentence will thus take $43 \log_2 27$ bits. The number of characters are recorded at $2 \log_2 43 \approx 10.85$ bits.[1] Altogether, the specification of the message requires $43 \log_2 27 + 2 \log_2 43 \approx 215.32$ bits.

However, in order to actually give a bound for Kolmogorov complexity, the length of the computer program which interprets the bits must also be included. Here is an example computer program in Python which could interpret the message

---

[1] A more compact representation for numbers is available. See the log^* method in Cover & Thomas (2006).

```
print ''.join(alphabet[index] for index in encoded_message)
```

Figure 7.5: An example Python program to interpret the encoded message

This assumes that the alphabet and encoded message are readily available and in a form amenable to processing within the language. It may be that the input has to be preprocessed, which would make the program longer. Additionally, the length of the program will vary heavily depending on which programming language is used. However, the distances between different computers and languages only differ by a constant (Cover & Thomas, 2006). As a result, it is common practice in algorithmic information theory to discount any actual program length and merely include that length as a constant, c. Consequently, the conditional Kolmogorov complexity can be expressed as

$$K(X|C) \leq 215.32 \text{ bits} + c. \tag{7.4}$$

The expression is less than rather than equal to because it is possible that an even more efficient way of expressing the sentence exists. However, at least this efficiency is possible.

The encoded version of the sentence requires 32 bits for each character, giving a total of 1,376 bits. Using a simplistic probability model, supposing that each bit is generated by the equivalent of a coin flip, the complexity, $-\log P(X)$, would be 1376 bits. Using equation 7.3,

$$A(X, C, p) = -\log(p) - K(X|C) \geq 1376 \text{ bits} - 215.32 \text{ bits} - c = 1160.68 \text{ bits} - c. \tag{7.5}$$

This shows 1,166 bits of algorithmic specified complexity by equation 7.3. Those 1166 bits are a measure of the confidence in rejecting the hypothesis that the sentence was generated by random coin flips. The large number of bits gives a good indication that it is highly unlikely that this sentence was generated by randomly choosing bits.

The hypothesis that the sentence was generated by choosing random English letters can also be analyzed. In this case the probability of this sentence can be calculated as

$$P(X) = \left(\frac{1}{27}\right)^{43}. \tag{7.6}$$

The complexity is then

$$-\log P(X) = -\log\left(\frac{1}{27}\right)^{43} = 43 \log 27 \approx 204.46 \text{ bits}, \tag{7.7}$$

in which case the algorithmic specified complexity becomes

$$A(X, C, p) = -\log p(X) - K(X|C) \geq 204.46 \text{ bits} - 215.32 \text{ bits} - c = -10.85 \text{ bits} - c. \tag{7.8}$$

The negative bound suggests no reason to suppose that this sentence could not have been generated by a random choice of English letters. The bound is negative as a result of two factors. In the specification, 10.85 bits were required to encode the length. On the other hand, the probability model assumes a length. Hence, the negative bits indicate information which the probability model had, but was not provided in the context. Since the only provided context is that of English letters, this is not a surprising result. No pattern beyond that explained by the probability model is identified.

The context can also be expanded. Instead of providing the English alphabet as the context, the word list of the *Oxford English Dictionary* can be used (OED Online, 2012). In the second edition of that dictionary there were 615,100 word forms defined or illustrated. For the purpose of the alphabet context, each letter is encoded as a number corresponding to that character. In this case, a number corresponding to words in the dictionary is chosen. Thus the number of bits required to encode the message using this context can be calculated:

$$K(X|C) \leq 9\log_2 615,100 + 2\log_2 9 + c \approx 179.41 + c. \tag{7.9}$$

Access to the context of the English dictionary allows much better compression than simply the English alphabet as comparing equations 7.4 and 7.9 shows.

Using equation 7.3 yields

$$A(X,C,p) = -\log p(X) - K(X|C) \geq 204.46 \text{ bits} - 179.41 \text{ bits} - c = 25.05 \text{ bits} - c. \tag{7.10}$$

This provides confidence to say this sentence was not generated by randomly choosing letters from the English alphabet.

It is possible to adopt a probability model that selected random words from the English language. Such a probability model would explain all of the specification in the sentence. It is also possible to include more information about the English language such that the specification would increase.

This technique depends on the fact that the numbers of words in the English language is much smaller then the number of possible combinations of letters. If the dictionary contained every possible combination of letters up to some finite length, it would not allow compression, and thus be of no help to finding evidence of specification. A language where all possible combinations of letters were valid words could still show specification, but another technique would have to be used to allow compression.

But one could also use a much smaller dictionary. A dictionary of 10 words would be sufficient to include all the words in this sentence. The ASC formula would give a much smaller compressed bound:

$$K(X|C) \leq 9\log_2 10 + 2\log_2 9 \approx 36.24 \text{ bits}. \tag{7.11}$$

This is a reduction of over 100 bits from equation 7.9. Because the sentence is much more closely related to the context, it takes about 16 bits less to encode each word when the dictionary is this small. In other words, it requires much less additional information to use the context when it is closely related to the message.

But it is possible to include words not included in the dictionary. The program would have to fall back on spelling the word one letter at a time. Only the bounds of the ASC can be computed. It is always possible a better compression exists, i.e., the object could be more specified than first realized.

3.2   Random Noise

While natural language is an example of something that should be specified, random noise is an example of something which should not. Consider a random bitstring containing 1,000 bits, where each bit is assigned with equal probability 1 or 0. Since randomness is incompressible, calculating the Kolmogorov complexity is easy. The only way of reproducing a random bitstring is to describe the whole bitstring.

$$K(X) \leq 2\log_2 1000 + 1000 + c \approx 1020 \text{ bits} + c \tag{7.12}$$

The probability of each bitstring is 2^{-1000}, and thus the complexity will be 1000 bits. Calculating the ASC:

$$A(X, C, p) = -\log p(X) - K(X|C) \geq 1000 \text{ bits} - 1020 \text{ bits} - c = -20 \text{ bits} - c. \tag{7.13}$$

As expected, the ASC is negative, and there is therefore no evidence of patterns in the string that are not explained by the probability model.

However, consider also the case of a biased distribution. That is, 1 and 0 are not equally likely. Instead, a given bit will be 1 two thirds of the time, while 0 only one third of the time. The entropy of each bit can be expressed as

$$H(X_i) = -\frac{1}{3}\log_2 \frac{1}{3} - \frac{2}{3}\log_2 \frac{2}{3} \approx 0.6365 \text{ bits} \tag{7.14}$$

for any i. The entropy of a bit is the number of bits required in an optimal encoding to encode each bit. This means the whole sequence can be described as

$$K(X) \leq 2\log_2 1000 + 1000 * H(X_i) + c \approx 656.5 \text{ bits} + c. \tag{7.15}$$

Using the uniform probability model, the complexity is still 1,000 bits and

$$A(X, C, p) = -\log p(X) - K(X|C) \geq 1000 \text{ bits} - 656.5 \text{ bits} - c = 343.3 \text{ bits} - c. \tag{7.16}$$

This random sequence has a high bound of algorithmic specified complexity. It is important to remember that the ASC bound only serves to measure the plausibility of the random model. It does not exclude the existence of another more accurate model that explains the data. In this case, using the actual probability model used to generate the message yields

$$-\log_2(p) = H(X_i) * 1000 \approx 636.5 \text{ bits} \tag{7.17}$$

and the resulting ASC:

$$A(X, C, p) = -\log p(X) - K(X|C) \geq 636.5 \text{ bits} - 656.5 \text{ bits} - c = -20 \text{ bits} - c.$$
(7.18)

The bound of ASC provides reason to reject a uniform noise explanation for this data, but not the biased coin distribution.

Dembski (Dembski, 1998) has considered the example of ballot rigging where a political party is almost always given the top billing on the ballot listing candidates. Since the selection is supposed to be chosen on the basis of a fair coin toss, this is suspicious. ASC can quantify this situation. The outcome can be described by giving the numbers of heads and tails, followed by the same representation as for the biased coin distribution.

$$K(X) \leq 2\log X_h + 2\log X_t + \log \binom{X_t + X_h}{X_h} + c$$
(7.19)

where X_h is the number of heads, X_t is the number of tails Assuming a probability model of a fair coin yields

$$-\log_2(p) = X_h + X_t \text{bits}.$$
(7.20)

This results in the following:

$$A(X, C, p) = X_h + X_t - 2\log X_h - 2\log X_t - \log \binom{X_t + X_h}{X_h} - c$$
$$= X_h + X_t - \log \left(X_h^2 X_t^2 \binom{X_t + X_h}{X_h} \right) - c.$$
(7.21)

Figure 7.6 shows the result of plotting this equation for varying numbers of head and tails given 20 coin tosses. As expected, for either high numbers of tails or high number of heads, the bound of ASC is high. However, for an instance which looks like a random sequence, the ASC is minimized.

3.3 Playing Cards

Another pertinent case is that of playing cards in poker. In playing cards, if the distribution is not uniform, somebody is likely cheating. For the purpose of investigating card hands, a uniform random distribution over all five-card poker hands is assumed.

In the game of poker, a poker hand is made up of 5 cards. Some categories of hands are rarer then others. Table 7.1 shows the frequency of the different hands.

Given a uniform distribution, every poker hand has the same probability and thus the same complexity. There are 2,598,960 possible poker hands. For a single hand, this yields a complexity of

$$-\log_2 p(X) = -\log_2(\frac{1}{2,598,960}) \approx 21.3 \text{ bits}.$$
(7.22)

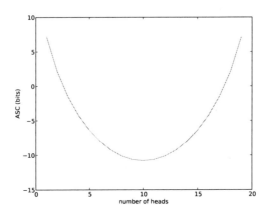

Figure 7.6: ASC for varyingly biased coin sequences and 20 coin tosses

Name	Frequency
Royal Flush	4
Straight Flush	36
Four of a Kind	624
Full House	3,744
Flush	5,108
Straight	10,200
Three of a Kind	54,912
Two Pair	123,552
One Pair	1,098,240
None	1,302,540

Table 7.1: Poker hand frequency

Name	Frequency	Complexity	Compressed Length	ASC
Royal Flush	4	21.310	5.322	15.988
Straight Flush	36	21.310	8.492	12.818
Four of a Kind	624	21.310	12.607	8.702
Full House	3,744	21.310	15.192	6.117
Flush	5,108	21.310	15.640	5.669
Straight	10,200	21.310	16.638	4.671
Three of a Kind	54,912	21.310	19.067	2.243
Two pair	123,552	21.310	20.237	1.073
One pair	1,098,240	21.310	21.310	0.000
None	1,302,540	21.310	21.310	0.000

Table 7.2: The ASC of the various poker card hands

While the probability of every poker hand is the same, the Kolmogorov complexity is not. To describe a royal flush requires specifying that it is a royal flush and which suit it is in. However, describing a pair requires specifying the paired value as well as both suits in addition to the three cards not involved in the pair. In general, describing a hand requires specifying the type of hand and which particular hand of all the possible hands of that type. This can be used to calculate the conditional Kolmogorov complexity for the hand.

$$K(H_i|C) \le \log_2 10 + \log_2 |H| + c. \tag{7.23}$$

where 10 is the number of types of hands. H is the set of all hands of a particular type, and H_i is a particular hand in that set.

There are 1,098,240 possible pairs. Putting this in Equation 7.23 gives:

$$K(H_i|C) \le \log_2 10 + \log_2 |H| + c \approx 23.39 \text{ bits} + c. \tag{7.24}$$

On the other hand, describing a pair without using the context gives

$$K(H_i|C) \le \log_2 2,598,960 + c \approx 21.3 \text{ bits} + c. \tag{7.25}$$

Single pairs are so common that the space required to record that it was a pair is more than the space required to record the duplicate card straightforwardly. Accordingly, the best approach is to take the minimum of the two methods

$$K(H_i|C) \le \min(\log_2 10 + \log_2 |H|, \log_2 2,598,960) + c. \tag{7.26}$$

Table 7.2 shows the ASC for the various poker hands. Rare hands have high ASC, but common hands have low ASC. This parallels expectations, because with a rare hand one might suspect cheating, but with a common hand one will not.

In other card games, a card is turned over after hands have been dealt to determine trump. The suit of the card is taken to trump for that round of the game. If the same suit is repeatedly chosen as trump, someone may ask what the odds are for that to occur. This question can be difficult to answer because every possible sequence of trump suits is equally likely. Yet, it is deemed unusual that the same suit is a trump repeatedly. Algorithmic specified complexity allows this to be modeled.

The suits are represented as a bit sequence using two bits for each suit,

$$K(X) = \log_2 4 + \log_2 H + c = 2 + \log_2 H + c \qquad (7.27)$$

where 4 is the number of suits, and H is the number of hands played. The complexity of the sequence is

$$-\log P(X) = 4^{\frac{-|X|}{2}} = 2H. \qquad (7.28)$$

The ASC is then

$$ASC(X, p) = 2H - 2 - \log_2 H - c. \qquad (7.29)$$

Note that this equation becomes $-c$ when $H = 1$. A pattern repeating once is no pattern at all and does not provide specification.

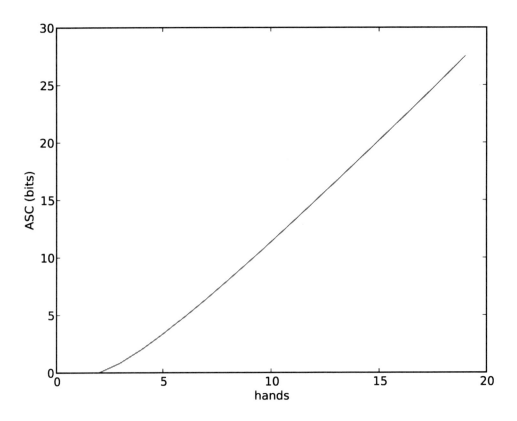

Figure 7.7: A plot of ASC for getting the same suit repeatedly

Figure 7.7 shows the ASC for increasing numbers of hands. The more times the same suit is chosen as trump, the larger the number of bits of ASC. The same trump for many rounds becomes less and less probable.

3.4 Folding Proteins

In biology, an important prerequisite to a protein being functional is that it folds. The fraction of all possible protein sequences that fold has been estimated: "the overall prevalence of sequences performing a specific function by any domain-sized fold may be as low as 1 in 10^{77}" (Axe, 2004).

A program can be created which uses the laws of physics to output a particular foldable protein.

```
for all proteins of length L
    run protein in a physics simulator
    if protein folds
        add to list of folding proteins
output the Xth protein from the list
```

Figure 7.8: A pseudocode program which uses a functional specification to compress the specification of a protein

Given different choices of L and N, this program will output any particular folding protein. This means that the protein can be described by providing those two numbers. Thus, the conditional Kolmogorov complexity can be calculated using these two numbers.

$$K(X|C) = 2\log_2 L + \log_2 F_L + c \qquad (7.30)$$

where C is the context, in this case the law of physics, and F_L is the number of folding proteins of length L. Taking Axe's estimate (Axe, 2004), and assuming simplistically that it applies for all lengths of proteins,

$$F_L = 10^{-77} 4^L \qquad (7.31)$$

$$\log F_L = -77\log 10 + L\log 4 \qquad (7.32)$$

therefore

$$K(X|C) = 2\log_2 L + \log_2 F_L + c = 2\log_2 L + -77\log 10 + L\log 4. \qquad (7.33)$$

The probability model will be chosen by supposing that each base along the DNA chain for the gene encoding the protein is uniformly chosen. It should be emphasized that according to the Darwinian model of evolution, the bases are not uniformly chosen. This supposition only serves to test a simplistic chance model of protein origin. The probability can be calculated as

$$- \log_2 \Pr(X) = - \log_2 4^{-L} = L \log_2 4. \tag{7.34}$$

Caution should be used in applying this formula. It assumes that the proportion of functional proteins is applicable for all lengths and implies that a fractional number of proteins fold.

Finally calculating the ASC,

$$\begin{aligned} ASC(X, p) &= L \log 4 - 2 \log_2 L + 77 \log_2 10 - L \log_2 4 - c \\ &= -2 \log_2 L + 77 \log_2 10 - c. \end{aligned} \tag{7.35}$$

The final bound for ASC depends little on the length of the protein sequence which only comes to play in the logarithmic term. The significant term is the $77 \log_2 10 \approx 255.79$ bits. Thus, there is good reason to believe that folding sequences were not generated randomly from a uniform distribution.

3.5   Functional Sequence Complexity

Kirk Durston et al. have defined the idea of *functional sequence complexity* (Durston, Chiu, Abel, & Trevors, 2007). Functional sequence complexity is related to a special case of algorithmic specified complexity.

A protein is made from a sequence of amino acids. Some sequences have functionality, and some do not. The case considered in section 3.4 above of folding is one particular case. Perhaps more interesting is considering the case of various proteins which perform useful biological functions.

Let Ω be the set of all proteins. Let F be the set of all proteins which pass a functionality test. Let $f(x)$ be a probability distribution over F. Both F and $f(x)$ can be produced by a simple algorithm using a functionality test on each element of Ω. Consequently, F and $f(x)$ can be described using a constant program length.

Consider the average for ASC over all elements in F.

$$\begin{aligned} \sum_{x \in F} f(x) A(x, C, p) &= \sum_{x \in F} f(x)(- \log p(x) - K(x|C)) \\ &= \sum_{x \in F} -f(x) \log p(x) - \sum_{x \in F} f(x) K(x|C)) \end{aligned} \tag{7.36}$$

Any element x can be described given the probability distribution and $\log f(x)$ bits. Given that $f(x)$ and F can be calculated with a constant program, the conditional Kolmogorov complexity can be calculated as

$$K(x|C) \leq \log -f(x) + c. \tag{7.37}$$

Place this into equation 7.36.

$$\sum_{x \in F} f(x) A(x, C, p) \geq \sum_{x \in F} -f(x) \log p(x) - \sum_{x \in F} -f(x) \log f(x) - \sum_{x \in F} c \qquad (7.38)$$

The middle term is recognized as the Shannon entropy.

$$\sum_{x \in F} f(x) A(x, C, p) \geq \sum_{x \in F} -f(x) \log p(x) - H(f) - c \sum_{x \in F} f(x) \qquad (7.39)$$

If the distribution p is uniform, $p(x) = \frac{1}{|\Omega|}$,

$$\sum_{x \in F} f(x) A(x, C, p) \geq \log_2 |\Omega| \sum_{x \in F} f(x) - H(f) - c \sum_{x \in F} f(x). \qquad (7.40)$$

The two summations over F are summations over a probability distribution and therefore 1.

$$\sum_{x \in F} f(x) A(x, C, p) \geq \log_2 |\Omega| - H(f) - c \qquad (7.41)$$

Equation 5 in Durston's work, adjusting for notation is

$$\log |\Omega| - H(f). \qquad (7.42)$$

This equation derives from making the same uniformity assumption made above. Thus, for the uniform probability distribution case,

$$\sum_{x \in F} (f(x) A(x, C, p)) + c \geq \log |\Omega| - H(f). \qquad (7.43)$$

This establishes the relationship between ASC and FSC. The difference is that the ASC is a lower bound and includes a constant. This is the same constant as elsewhere: the length of the program required to describe the specification.

4 Objections

4.1 Natural Law

It has been argued in this paper that compressibility in the presence of context is a necessary condition for information. This is in contrast to others who have argued that lack of compressibility is a necessary condition for information (Abel & Trevors, 2005). But compressible objects lack complexity. Because a compressible object is describable as some simple pattern, it is amenable to being produced by a simple process. Many objects in the real world follow simple patterns. Water tends to collect at lower elevations. Beaches follow a sloping pattern. Sparks fly upwards. But these patterns are the result of the operation of simple law-like processes. Even if the explanations for these patterns were unknown, the simplicity of the pattern suggests that some simple explanation existed.

The premise behind this use of compressibility is that it identifies what human would see as simple patterns. Abel writes: "A sequence is compressible because it contains redundant order and patterns" (Abel & Trevors, 2005).

The problem is that algorithms are very versatile and allow the description of many patterns beyond that which humans would see as patterns. As has been shown by the various examples in this paper, many objects which do not exhibit what humans typically identify as redundant order and patterns are in fact compressible. Significantly, functionality actually allows compressibility. Contrary to what Abel states, functional sequences are compressible by virtue of the functionality they exhibit. All of the sequences that Abel holds to be mostly incompressible are actually compressible.

But are compressible objects amenable to explanation by simple processes? Do all compressible objects lack complexity? If this were true, it would be problematic for algorithmic specified complexity because all specified objects would also not be complex, and no object would ever be both specified and complex. But many compressible objects do not appear amenable to explanation by a simple process.

As discussed, English text is compressible given a knowledge of the English language. This does not somehow make it probable that English text will appear on a beach carved out by waves. Ninety degree angles are very compressible; yet, they are not typically found in nature. The existence of an explanation from the laws of nature does not appear to follow from compressibility.

Kolmogorov complexity deliberately ignores how long a program takes to run. It is only concerned with the length of the program's description. A program may be short but take an astronomical amount of time to run. Many of the specifications considered in this paper fall into that category. These objects are compressible, but that compression does not give a practical way to reproduce the object. But if there is no practical way to reproduce the object, there is no reason to suggest law-like processes as a plausible explanation.

4.2   Context is Subjective

The ASC of any object will depend on the context chosen. Any object can be made to have high ASC by using a specifically chosen context. But this appears to be the way that information works. If the authors, who do not understand Arabic, look at Arabic text, it appears to be no better then scribbling. The problem is not that Arabic lacks information content, but that the reader is unable to identify it without the necessary context. As a result, this subjectivity appears to capture something about the way information works in the human experience.

As with specification, it is important that the context be chosen that is independent of the object under investigation. While a specification will rarely be independent of the object under investigation, it is much easier to maintain this independence in the case of a context.

4.3 Incalculability

It is not possible to calculate the Kolmogorov complexity of an object. However, it is possible to upper-bound the Kolmogorov complexity and thus lower-bound the algorithmic specified complexity. This means that something can be determined to be at least this specified, although the possibility that it is even more specified cannot be ruled out. Therefore, even though detecting a specification cannot be achieved mechanically, it can be objectively identified when found.

Acknowledgements

The approach of using compressibility as a measurement of specification was suggested to the authors by Eric Holloway. The authors have attempted to extend the approach to apply to many more types of specifications. The authors are grateful for his initial suggestion and answer to our initial objections to the idea.

References

Abel, D. L. & Trevors, J. T. (2005). Three subsets of sequence complexity and their relevance to biopolymeric information. *Theoretical Biology & Medical Modelling*, 2, 29. Available from http://www.pubmedcentral.nih.gov/articlerender.fcgi?artid=1208958&tool=pmcentrez&rendertype=abstract, doi:10.1186/1742-4682-2-29

Axe, D. D. (2004). Estimating the prevalence of protein sequences adopting functional enzyme folds. *Journal of Molecular Biology*, 341(5), 1295-315. Available from http://www.ncbi.nlm.nih.gov/pubmed/15321723, doi:10.1016/j.jmb.2004.06.058

Bennett, C. H. (1988). Logical depth and physical complexity. In *The universal turing machine: A half-century survey* (pp. 227-257). Available from http://www.springerlink.com/index/HRG11848P291274Q.pdf

Chaitin, G. J. (1966). On the length of programs for computing finite binary sequences. *Journal of the ACM (JACM)*, 13. Available from http://dl.acm.org/citation.cfm?id=321363

Chaitin, G. J. (2002). *Conversations with a mathematician: Math, art, science, and the limits of reason: A collection of his most wide-ranging and non-technical lectures and interviews.* New York: Springer.

Cover, T. M. & Thomas, J. A. (2006). *Elements of information theory.* Hoboken, NJ: Wiley-Interscience, second edition.

Dembski, W. A. (1998). *The design inference: Eliminating chance through small probabilities.* Cambridge University Press. Available from http://mind.oxfordjournals.org, doi:10.1093/mind/112.447.521

Dembski, W. A. (2002). *No free lunch: Why specified complexity cannot be purchased without intelligence.* Lanham MD: Rowman & Littlefield. Available from http://www.worldcat.org/title/no-free-lunch-why-specified-complexity-cannot-be-purchased-without-intelligence/oclc/46858256&referer=brief_results

Dembski, W. A. (2005). Specification: The pattern that signifies intelligence. *Philosophia Christi*, 7(2), 299-343. Available from http://www.lastseminary.com/specified-complexity/Specification-ThePatternThatSignifiesIntelligence.pdf

Durston, K. K., Chiu, D. K. Y., Abel, D. L., & Trevors, J. T. (2007). Measuring the functional sequence complexity of proteins. *Theoretical Biology & Medical Modelling*, 4, 47. doi:10.1186/1742-4682-4-47

Kolmogorov, A. N. (1968a). Logical basis for information theory and probability theory. *IEEE Transactions on Information Theory*, 14(5), 662–664. Available from http://ieeexplore.ieee.org/lpdocs/epic03/wrapper.htm?arnumber=1054210, doi:10.1109/TIT.1968.1054210

Kolmogorov, A. N. (1968b). Three approaches to the quantitative definition of information. *International Journal of Computer Mathematics*. Available from http://www.tandfonline.com/doi/abs/10.1080/00207166808803030

Levin, L. A. (1976). Various measures of complexity for finite objects. (axiomatic description). *Soviet Math*, 17(522).

OED Online (2012). Oxford English dictionary online. Available from http://dictionary.oed.com

Shannon, C. E., Weaver, W., & Wiener, N. (1950). The mathematical theory of communication. *Physics Today*, 3(9), 31. Available from http://www.ncbi.nlm.nih.gov/pubmed/9230594, doi:10.1063/1.3067010

Solomonoff, R. J. (1960). *A preliminary report on a general theory of inductive inference*. Technical report, Zator Co. and Air Force Office of Scientific Research, Cambridge, Mass. Available from http://citeseerx.ist.psu.edu/viewdoc/download?doi=10.1.1.66.3038&rep=rep1&type=pdf, doi:10.1.1.66.3038

8

Complex Specified Information (CSI) Collecting

ERIC HOLLOWAY

Independent Scholar

Abstract

Intelligent Design Theory makes use of the concept of intelligent agency as a distinct causal mode, distinct from chance or algorithmic modes of causation. The influence of such agency is often detected by the contribution of information to a search process. Assuming humans are capable of such causal roles, then it should be possible to measure the amount of information that a human is contributing to such a process. This is done by measuring the success rate for a search for a solution to a computationally hard problem by both humans and computers. The methodology used for this experiment was not successful, but it is hoped that the experimental setup and methodology will inspire further improvement and research in this area.

1 Introduction

Research comparing human cognitive capabilities to computer algorithms suggest humans possess supra-computational cognition. Humans are capable of finding good solutions to computationally intractable problems, and this capability scales at a faster rate with larger problems than the best known algorithms (Dry, Lee, Vickers, & Hughes, 2006). Many popular games are algorithmically intractable to solve (Viglietta, 2011). Programmers appear capable of solving halting problems (Bartlett, 2014a). People can solve insight problems, for which there is currently no known computational method (Bartlett, 2014b). Observers can pick out targets in a picture relevant to their goal independently of the target's features (Gue, Preston, Das, Giesbrecht, & Eckstein, 2012). Such capabilities either have never been algorithmically emulated, or violate well known and well substantiated computational constraints, such as the computational complexity of the problem (Cormen, Leiserson, Rivest, & Stein, 2001), the No Free Lunch Theorem (NFLT) (Wolpert & Macready, 1996, 1995) and the halting problem (Cover & Thomas, 2006).

Intelligent Design Theory (IDT) provides a formal language to account for observations of supra-computational cognition. According to IDT, intelligent agents, such as humans, are capable of creating a new form of information called complex specified information (CSI) (Dembski, 2005). Dembski's work on search algorithms (Dembski, 2006) implies that if human interaction can be incorporated into a search and optimization algorithm, these algorithms can surpass the limitations of the NFLT over a wide variety of hard problems. This is possible due to the unique ability of intelligent agents, such as humans, to create CSI. A further implication is the user does not need specialized interfaces or training for each type of problem. The possibility of such a generalized interface is suggested by research demonstrating context independent problem solving by humans (Sperber, Cara, & Girotto, 1995) and suggested by IDT's implications.

2   Problem Description

In the computational domain, it is problematic to identify whether human interactions can be defined by an algorithm. Since all the interactions with the computer are defined by a series of 1s and 0s, i.e., a bit string, the interaction can be codified by an algorithm that outputs the same bit string. Consequently, on first analysis, trying to computationally distinguish human interaction from algorithm output seems impossible. If human interaction is indistinguishable from algorithmic output, then it is not plausible human interaction can surpass the No Free Lunch Theorem (NFLT). Such is the basic problem addressed: How is it possible to distinguish human interaction from algorithm output?

Even though it is always possible to codify a series of events in a finite domain after the fact, the question is whether such codification is possible prior to the event's occurrence. A prior codification is not possible in all cases; otherwise, it violates the halting problem (Cover & Thomas, 2006). This means the possibility of codifying a human's interaction after the interaction has occurred does not necessarily entail it can be codified before the interaction. Consequently, it is possible that with the right experimental design a human's interaction can be distinguished from an algorithm output.

3   Background

The question is, what is the experimental design that can make the distinction between human and algorithm? One approach is to use the NFLT. The theorem sets precise, rigorous boundaries on the mathematical capabilities of algorithms. Anything performing outside of these boundaries is by logical necessity non-algorithmic. To show humans are supra-computational, one need simply show they do not abide by the NFLT. Unfortunately, such a demonstration may be impossible. The NFLT only applies across all problems within a very small, specialized subset of problem types. As such, the NFLT is quite difficult to apply in practice.

The Almost No Free Lunch Theorem (ANFLT) (Droste, Jensen, & Wegener, 2002) provides a solution. The theorem shows while the original NFLT does not apply to most particular problem domains, the expected performance for most algorithms on many real world problems does not vary significantly between algorithms. There is almost no free lunch for a large portion of real world problems.

The ANFLT result implies that given a search algorithm, as long as the algorithm is selected independently of the problem, it is unlikely the choice of algorithm makes a significant difference in search performance. Consequently, if human interaction discovers a solution significantly better than search algorithms, it is very likely the interaction was non-algorithmic, and therefore supra-computational.[1]

According to William Dembski's research on search algorithms (Dembski & Marks II, 2010), this supra-computational interaction takes the form of active information creation. Active information is the information necessary to improve a search algorithm's expected speed in finding a target solution beyond that of a random search. That is, with more active information, fewer search queries are necessary to find the target solution.

Active information can be inserted into the algorithm from an existing source of external information, or it may be created. In the case of the ANFLT, if the search algorithm exhibits significant amounts of active information (i.e., finds a target much faster than statistically expected), this information must be created since the conditions of the ANFLT prohibit the information from being merely transferred from an external source.

Since the information is created, it cannot come from either chance or necessity, as these sources can only degrade or transfer already existing information. Additionally, the information is specified by the degree it reduces the number of search queries to find the target solution. Information that is neither the product of chance nor necessity and is specified is complex specified information (CSI) (Dembski, 2005) as defined by Intelligent Design Theory (IDT). Furthermore, IDT claims CSI can come from intelligent agents.

---

[1]Throughout this paper, the term "solution" refers to both the best/true solution to a problem, as well as substandard solutions to a problem. Even in the case where there is only one true answer to a problem, there may be other answers that are approximations of the true answer. The two types of solutions are distinguished in terms of their optimality. The true solution is the optimal solution.

4   Approach

The approach in this study is to develop and test a general technique for integrating human interaction into search and optimization algorithms. By incorporating human interaction in this way, it is possible to determine whether humans violate the algorithmic search constraints of the ANFLT and consequently create CSI as predicted by IDT. This technique's implementation consists of a software framework that can be integrated with many different kinds of algorithms and problems, including both single and multi-objective problems. The technique is referred to as CSI Collecting (CSIC) throughout the rest of this paper.

CSIC is demonstrated as a proof of concept by using it to find public and private keys for the RSA asymmetric cryptography system (Cormen et al., 2001). The algorithm used in CSIC is a multi-objective genetic algorithm (Coello, Lamont, & Veldhuizen, 2007). The human users will use CSIC through Amazon's Mechanical Turk service.

Success in the experiment is measured by the rate of solution improvement (fitness increase) compared with problem information discovered, which is Fitness/Information or FI for short. Solution improvement is measured by an objective function in the genetic algorithm. The amount of problem information discovered is measured by the number of solutions evaluated. The improvement due to human interaction is compared to the improvement due to the genetic algorithm as measured by FI.

Since the project is currently in the exploratory stage, the comparison is informal. It is assumed the algorithm discovers problem domain information at an exponentially greater rate compared to human agents. Consequently, any greater or equivalent improvement of fitness by the human agents compared to the algorithm produces a very high FI value in favor of the human agents. Thus, the comparison between human agents and the genetic algorithm is based purely on fitness increase.

5   Implementation

The general technique used is as follows. A standard multi-objective genetic algorithm is used, a type of stochastic global search algorithm (Figure 8.1). A genetic algorithm takes a set of solutions, measures how good each solution is according to some fitness valuation function, varies the solutions to generate a new set using variation operators such as mutation and crossover, and then from both sets of solutions selects an output set according to some criteria. The algorithm then reiterates this process on each subsequent output set until a stopping criteria is reached. The solutions themselves are represented to the algorithm as fixed length bit strings. The functions *generate* and *select* in Figure 8.1 can each incorporate an intelligent agent. In the implemented version of CSIC, only the *select* function incorporates human interaction, even though the more idealized version can incorporate human interaction in the *generate* phase

$\mathcal{S}_i^p \leftarrow \{init()\};$
$\mathcal{S}_o^f[0] \leftarrow \{\};$
$t := 0;$
Ensure: $\forall s^p \{s^p \in \mathcal{S}[t] \rightarrow s^p \in \mathcal{S}_i^p\}$
Ensure: $\forall s^f \{s^f \in \mathcal{S}^f[t] \rightarrow s^f \in \mathcal{S}[t] \wedge \mathcal{F}(s^f)\}$
Ensure: $\forall s^f \{s^f \in \mathcal{S}_o^f[t] \rightarrow s^f \in \mathcal{S}^f[t] \wedge \mathcal{C}(s^f, \mathcal{S}_o^f[t-1])\}$
   while $\mathcal{O}(\mathcal{S}_o^f[t], t) \neq TRUE$ **do**
      $t := t + 1;$
      $\mathcal{S}[t] := generate(\mathcal{S}_i^p, \mathcal{S}[t-1])$
      $\mathcal{S}^f[t] := feasible(\mathcal{S}[t])$
      $\mathcal{S}_o^f[t] := select(\mathcal{S}^f[t], \mathcal{S}_o^f[t-1])$
   end while

p	partial solution
f	feasible solution
i	input set
o	output set
$\mathcal{S}$	set of solutions, can be either partial or feasible
$\mathcal{F}$	feasibility, function of solution feasibility, returning true or false
$\mathcal{C}$	comparison, whether solution is selected when compared to solution set
$\mathcal{O}$	objective, function of both output solution set and current iteration count

Figure 8.1: Stochastic global solution search algorithm

of the algorithmic process.

5.1 Data Flow

The data flow between the human and the algorithm is demonstrated in Figure 8.2. Both the unguided and guided algorithms are run in tandem on the same solution set for greater effectiveness in discovering new solutions. For experimental purposes, this does muddy the data to an extent, but not irrevocably. Human-provided solutions are uniquely identified, and the phylogeny of each solution is tracked, making it possible to derive the impact of human interaction.

The two processes are combined into a hybrid system because for real world use algorithms and humans work well together. The algorithmic side is very good at checking many solutions very quickly, thereby providing the user with better information for making decisions on new areas of the problem to explore.

5.2 User Interface

In the user interface, the user is presented with a list of solutions from which he can make his selection. The solutions are the strings of arbitrary symbols in Figure 8.3.

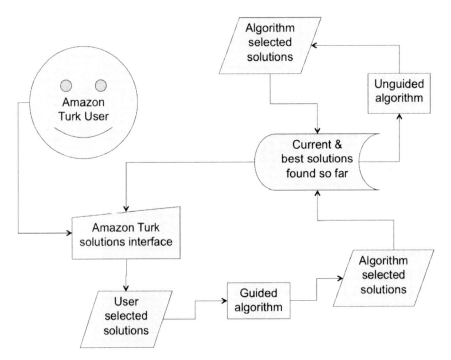

Figure 8.2: CSIC data flow between user and algorithm.

Each solution is represented symbolically, so the user has no contextual clues as to what the problem is that he is attempting to solve. This is done in order to create a context independent interface that can be used for a wide variety of problems and algorithms. The only information to which the user has access is similarities between solutions (as shown by similarities in symbols) and the value of each solution. Once the user selects a set of solutions, he picks the type of variation operators to apply to the solutions and then presses "GO" to have the algorithm, as shown in Figure 8.1, create a new set of solutions. Dembski's work implies that even with only this information it is possible for the user to be able to provide active information to the search algorithm because a prior external source of information is unnecessary to add active information to the search since the user is an intelligent agent.

The types of operators have different credit costs since some operators discover more information about the problem than others. It is important to track the amount of information with which the user is making his decisions in order to fairly compare his performance to that of the unguided algorithm. The user is rewarded based on whether this new set of solutions improves over the best solutions found so far.

The actual implementation in Figure 8.4 is a simplified version of the interface shown in Figure 8.3, since it is used by people on the Amazon Turk service. Due to the low wage for using the interface, they cannot be expected to spend a long time trying to understand the intricacies of different variation operators. Accordingly, credits are done away with since the same operators are used in every iteration. Additionally, the numerical score is replaced with a visual indication of solution value, where higher

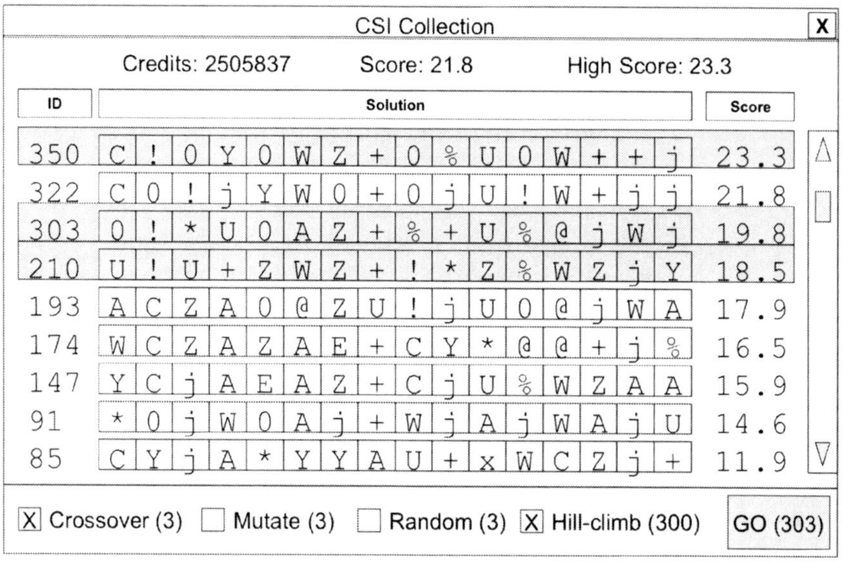

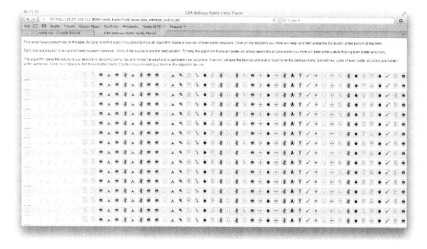

Figure 8.3: User interface concept for CSIC.

Figure 8.4: Actual user interface website.

valued solutions have a greater number of stars.

Otherwise, the basic idea is the same between the actual and conceptual interfaces. The users select solutions to guide the search algorithm and then press "GO." The users are paid a basic wage for using the webpage and are rewarded with a bonus based on the value of new solutions discovered. The intent is to financially incentivize the users to discover highly valued and highly unique solutions.

5.3   Problem

CSIC is used to find the prime factors that generate keys for an asymmetric cryptosystem. There is no known correlation between finding prime factors and multi-objective genetic algorithms, nor with human input during the selection phase. This means the problem and algorithm together meet the criteria for the ANFLT to apply.

Breaking asymmetric encryption is also an extremely hard problem, which is why it forms the basis of much information transmission security. The asymmetric cryptosystem is RSA (Cormen et al., 2001). The genetic algorithm uses a multi-objective fitness function to measure how close a solution is to the correct set of primes. To find the primes, the fitness function is given access to a true plaintext and true cyphertext. The two objectives are:

1. *Matching encrypted bits.* The plaintext is encrypted using the keys generated by the solution primes. This encrypted text is then compared with the true cyphertext, and the number of matching 1 bits are counted. The 0 bits are not counted because they generally overwhelm the number of 1 bits.

2. *Matching decrypted bits.* The true cyphertext is decrypted using the solution keys, and its accuracy is measured as for the previous objective.

6   Results and Conclusion

The guided genetic algorithm received 500 human inputs from the Amazon Mechanical Turk service. The human input contributed 1 out of 18 superior solutions found (Figure 8.5). The one solution found by a human was also found by the algorithm. However, the logs show all human input consisted of selecting every single solution, which can be replicated by an algorithm.

This result did not validate the hypothesis, though the failure is due to methodological factors, as will be discussed in Section 8. Since both the human and algorithm discovered the same solution, it is not clear the solution was originally found by the human. Thus, in terms of the FI metric in Section 4, it is not possible to say whether the human outperformed the algorithm.

7   Theoretical Objections

While the methodology for CSIC is still in its infancy, and its efficacy has yet to be demonstrated, there are also theoretical objections to the concept that have been or could be raised. The following section attempts to provide answers to these objections. In all of these objections the assumption is made that CSI, or active information, cannot be created by algorithmic sources. Additionally, it is also assumed that active information is a subcategory of CSI.

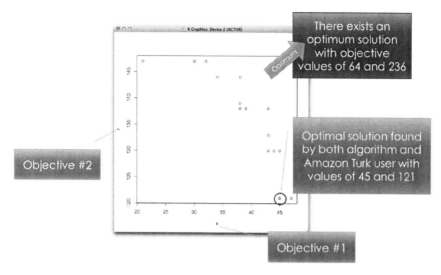

Figure 8.5: Best solutions found from Amazon Turk experiment. The objectives are described in Section 5.3.

7.1 Objection: Human intelligence is an algorithm with already high amounts of CSI

One objection to this methodology is that human intelligence is an algorithm that *already contains* high amounts of CSI, especially since the ANFLT does not rule out the possibility that a proximate algorithmic information source is available. It might be the case that human minds are highly tuned algorithms for particular difficult problem domains. Consequently, humans will outperform all current state of the art algorithms in these domains, but this does not preclude human intelligence being algorithmic. This objection holds even in the case of the ANFLT.

Such an objection is valid. The CSIC experiment cannot categorically rule out the possibility that human intelligence is a kind of algorithm. However, CSIC can provide an inference to the best explanation between two competing hypotheses. The first hypothesis asserts that the proximate cause, human intelligence, created the information, while the second hypothesis states that a more remote intelligent agent created the information.

To determine which hypothesis is better supported by a positive result in the CSIC experiment, the principle of minimal CSI creation is proposed. The principle is analogous to Ockham's Razor. This principle states in the case where the creation of CSI is presupposed, the explanation relying on the least amount of CSI creation is preferred.

The first hypothesis relies on less CSI creation since the human only creates CSI for a particular problem instance. Hypothesis two requires the creation of enormous amounts of CSI covering all possible problem instances the human might encounter.

Consequently, while the CSIC experiment does not rule out algorithmic human intelligence, it does show that supra-computational human intelligence is the best explanation of a positive result.

7.2 Objection: Supra-computation does not imply effectiveness on all problems

Another objection to this methodology is that even if human intelligence is supra-computational, this does not imply that it is necessarily effective on all possible problems. It could be performing supra-computational actions which are still limited in scope and thus preventing its use in arbitrary problems. This objection is also valid. Consequently, the question is, in which problems do humans demonstrate supra-computational abilities? One way to answer this question is experimentally, by having humans work with different problems and seeing in which instances supra-computational abilities are exhibited. This is the intent of the CSIC experiment. There is no presumption supra-computational abilities will be demonstrated, only a presumption that it is possible to detect such abilities within the experiment.

Research by Bartlett (2014b) suggests that even if humans do demonstrate supra-computation, such a capability may depend on existing information, such as axioms about the problem domain. If true, then the almost complete removal of context in CSIC may render the user incapable of contributing active information to the search algorithm.

7.3 Objection: Supra-computation is only exhibited in context dependent interactions

This objection states that the removal of almost all context in CSIC is a methodological problem because experience seems to indicate that supra-computation only occurs within situational contexts. A solution can only be created when the problem is contextually understood.

However, the observation is not always true. One scenario similar to CSIC where information is created is learning to read. When people learn to read a language, they are presented with a string of symbols without an inherent context. They learn the meaning of the symbols through external responses, such as getting affirmation when mapping the symbols to the correct action or object.

7.4 Objection: Supra-computation requires holistic reasoning

This objection states that in order to solve a problem, the person solving must not only know the problem, but also understand why the problem exists. Since the user of CSIC neither knows nor understands the problem at hand, he will be unable to solve the problem.

There is a degree of holism available in CSIC, as the user can see many solutions and valuations and thus look for overarching patterns. The user can understand at a very abstract level the characteristics of good and bad solutions.

7.5 Objection: Fitness function in experiment cannot factor primes for RSA keys

This objection is quite likely correct. However, even if correct, if human interaction improves the solutions beyond the algorithmic limits, then the experiment achieves a positive result. On the other hand, better choices of problems, such as well understood pedagogical problems, would greatly improve the experimental design.

RSA cracking was chosen for this experimental investigation mostly because the applicability of solving this problem is much easier to explain to a lay audience than more pedagogical problems. Additionally, a successful result for this problem would have direct, groundbreaking relevance for software engineering.

8 Future Work

While the essential methodology and implementation of CSIC have now been created and tested, many areas of improvement remain. Additionally CSIC must be compared to alternatives to see if CSIC is truly effective and beneficial.

The main potential area of improvement is in verifying whether the Amazon Turk input is truly human generated and whether the users are actually trying to find patterns. Users have been known to script Amazon Turk jobs, and without such verification in this case it is not possible to know with certainty whether the input is human generated. When the logs from this experiment were analyzed, it turns out the Amazon Turk users were just selecting all the solutions and not trying to find patterns in the solutions. To provide a cleaner environment for experimentation, the guided and unguided algorithms should be separated. Additionally, the data logging needs to be time-stamped.

The algorithm and problem used in the experiment can be improved in numerous ways. Pedagogical problems and algorithms should be used to compare the effectiveness of CSIC to the current state of the art. Practical problems where search algorithms have already proven effective should be explored to see if CSIC can provide additional benefit. CSIC should also be compared to contextualized human-powered search algorithms, such as Foldit, to see how the addition of context affects search effectiveness.

Different forms of motivation should be explored. The Amazon Turk interface relies on a financial motivation, which motivates users to cut corners and perform the task in the fastest way possible. Such motivation does not encourage users to find extremely good solutions. If the interface is in the form of an entertaining game, users are better encouraged to find good solutions. Furthermore, a simplified explanation of the cutting edge relevance of their work provides an intrinsic motivation, and encourages innovation as demonstrated by significant user innovation in Foldit (Moore, 2012).

References

Bartlett, J. (2014a). Calculating software complexity using the halting problem. In J. Bartlett, D. Halsmer, & M. R. Hall (Eds.), *Engineering and the ultimate* (pp. 123–130). Broken Arrow, OK: Blyth Institute Press.

Bartlett, J. (2014b). Using Turing oracles in cognitive models of problem-solving. In J. Bartlett, D. Halsmer, & M. R. Hall (Eds.), *Engineering and the ultimate* (pp. 99–122). Broken Arrow, OK: Blyth Institute Press.

Coello, C. A. C., Lamont, G. B., & Veldhuizen, D. A. V. (2007). *Evolutionary algorithms for solving multi-objective problems*, chapter MOEA Parallelization. Springer: New York.

Cormen, T. H., Leiserson, C. E., Rivest, R. L., & Stein, C. (2001). *Introduction to algorithms*. MIT Press, second edition.

Cover, T. M. & Thomas, J. A. (2006). *Elements of information theory*. Wiley-Interscience, 2nd edition.

Dembski, W. (2005). Specification: The pattern that specifies intelligence. Available from http://www.designinference.com/documents/2005.06.Specification.pdf

Dembski, W. (2006). *Conservation of information in search*. Technical report, Center for Informatics.

Dembski, W. A. & Marks II, R. J. (2010). The search for a search: Measuring the information cost of higher level search. *Journal of Advanced Computational Intelligence and Intelligent Informatics*, 14(5), 475–486.

Droste, S., Jansen, T., & Wegener, I. (2002). Optimization with randomized search heuristics: the (a)nfl theorem, realistic scenarios, and difficult functions. *Theoretical Computer Science*, 287, 131–144.

Dry, M., Lee, M. D., Vickers, D., & Hughes, P. (2006). Human performance on visually presented traveling salesperson problems with varying numbers of nodes. *The Journal of Problem Solving*, 1(1), 20 – 32.

Gue, F., Preston, T. J., Das, K., Giesbrecht, B., & Eckstein, M. P. (2012). Feature-independent neural coding of target detection during search of natural scenes. *The Journal of Neuroscience*, 32(28), 9499–9510.

Moore, E. A. (2012). Foldit game leads to aids research breakthrough. Available from http://news.cnet.com/8301-27083_3-20108365-247/foldit-game-leads-to-aids-research-breakthrough/

Sperber, D., Cara, F., & Girotto, V. (1995). Relevance theory explains the selection task. *Cognition*, 57, 31–95.

Viglietta, G. (2011). Gaming is a hard job, but someone has to do it! *arXiv*. Available from http://arxiv.org/pdf/1201.4995.pdf

Wolpert, D. H. & Macready, W. G. (1995). *No free lunch theorems for search*. Technical report, Santa Fe Institute.

Wolpert, D. H. & Macready, W. G. (1996). *No free lunch theorems for optimization*. Technical report, Santa Fe Institute and TXN Inc.

Part IV

The Engineering of Life

Over the past several decades, engineering and biology have become more and more integrated with each other. Engineering often supplies conceptual frameworks to understand biological systems (sometimes termed "systems biology"), as well as an understanding of engineering problems being solved by living organisms. Likewise, in the emerging field of biomimetics, biology serves as an inspiration to engineers, who are increasingly incorporating designs and technical solutions inspired by biological organisms into their work. In this part of the book, Arminius Mignea examines the biological process of self-replication from an engineer's perspective in a three-part series. The first part takes a systems biology approach and looks at the large-scale design requirements of any successful physical self-replicator. The second part then considers a more biomimetic question of how such a physical replicator might be implemented using current technology. The final part of the series reflects upon the potential implications of the existence of biological self-replication to origin-of-life scenarios.

9 || Developing Insights into the Design of the Simplest Self-Replicator and Its Complexity: Part 1—Developing a Functional Model for the Simplest Self-Replicator

ARMINIUS MIGNEA

The Lone Pine Software

Abstract

This is the first in a three-part series investigating the internals of the simplest possible self-replicator (SSR). The SSR is defined as having an enclosure with input and output gateways and having the ability to create an exact replica of itself by ingesting and processing materials from its environment. This first part takes an analytical approach and identifies, one by one, the internal functions that must operate inside the SSR to be a fully autonomous replicator.

1 Introduction

One of the most remarkable characteristics of living organisms is their ability to self-replicate. There are many forms and manifestations of self-replication. These forms vary from the simplest, single-celled organisms to a wide range of living forms to the most complex organisms, including humans and other mammals.

One of the most intriguing questions that ordinary people, engineers, scientists, and philosophers have obsessed over for centuries is how life on Earth originated and is able to create descendants that look like their parents. Many researchers and scientists have invested tremendous resources in trying to identify a plausible natural means by which the simplest forms of life may have been created from inanimate matter. They have tried to identify, and hopefully reproduce, a set of events and circumstances that somehow puts together the basic elements of the simplest entity to replicate and thus become a living organism.

The goal of this study is to use an engineering approach to develop insights into the internal design of a simplest possible self-replicator (SSR). The SSR is defined for the purpose of this study as an autonomous artifact that has the ability to obtain material input from its environment, grow, and create an exact replica of itself. The replica should "inherit" the ability from its "mother" SSR to create, in its turn, an exact copy of itself.

It is important to observe that this simple definition of the SSR accurately mimics the characteristic behavior of many single-celled organisms, at least from the perspective of their ability to self-reproduce. In particular, they are autonomous in regards to their ability to ingest materials from their environment, to use the ingested materials for growth and production of internal energy, and to produce an identical copy of themselves, usually through a two-step process of cloning and division.

Part 1 and Part 2 of this paper will analyze the process of self-replication as though it is a preliminary study for a research lab tasked to design and build an artificial SSR from scratch. The objectives of this study are as follows:

1. Create a top level design of the SSR that identifies all its functional components with specific characterization of the role of each functional component, its responsibilities and its interactions with the other functional components within the SSR framework.

2. Identify the candidate engineering technologies that could be employed in the concrete implementation of the SSR and of its functional components.

3. Identify the most difficult tasks in constructing the concrete artificial SSR.

4. Conclude with an overall estimate of the complexity in constructing the artificial SSR. For a pragmatic estimation of SSR construction complexity, it may be compared with existing, high technology, human-made artifacts.

Because this is a thought experiment on what elements would be required to construct a simple self-replicator and makes use of general knowledge in the field of engineering, references have not been included at the end of part one.

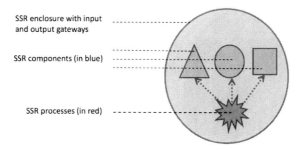

Figure 9.1: The SSR Structure

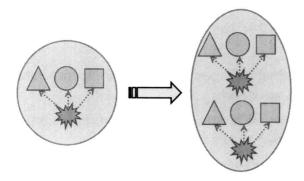

Figure 9.2: The Cloning Phase in SSR Replication

2   The Two Phases of Self-Replication

At the highest level, the SSR has the composition illustrated in figure 9.1.

The SSR replication process has two main phases—the cloning phase (illustrated in figure 9.2) and the division phase (illustrated in figure 9.3).

The behavior of the SSR, including basic support functions and the two replication phases, can be outlined as follows:

- Input raw materials and raw parts are accepted by input enclosure gates.

- Input raw materials are processed through material extraction into good materials for fabrication of parts or for energy generation.

- Energy is generated and made available throughout the SSR.

- Processed materials are passed to the fabrication function which fabricate parts, components and assemblies for cloning of all SSR internal elements, creating scaffolding elements for the growing SSR interior, and creating new elements that are added to the growing enclosure.

- When the cloning of all original SSR internal parts is completed, the SSR division starts:

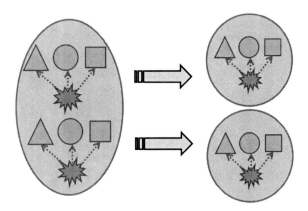

Figure 9.3: The Division Phase in SSR Replication

- The original SSR contents are now at, for example, the north pole of the SSR enclosure.

- The cloned SSR contents (the nascent daughter SSR) are now at the "south pole" of the SSR enclosure.

- The SSR enclosure and its content now divide at the equatorial plane and the separate mother SSR (at north) and daughter SSR (at south) emerge.

3 Identifying SSR Capabilities as Specific Functions

By conducting a step-by-step analysis of what must be happening inside and at the periphery of the SSR, its growth and replication abilities can be characterized.

3.1 The SSR enclosure and its input and output gateways

Two reasonable assumptions will be made about the SSR. The first assumption is that the SSR is comprised of an enclosure that has the role of separating the SSR from its environment. Secondly, the surface of this enclosure has openings that are used to accept good substances into the SSR—raw materials and raw parts from the SSR environment. These openings will be called *input gateways*. There are also openings used by the SSR to expel from inside the SSR refuse materials and parts that result from certain transformation/fabrication processes. These openings will be called *output gateways*.

3.2   The input flow control function

There is one primary question for the input gateways: Are all the raw materials and raw parts that exist or touch the outside of the enclosure good for the SSR processes? Certainly, they are not. The SSR and its input gateways must feature some ability to select or reject substances, materials, and parts that are outside the SSR interior and determine whether they should enter. For this paper, this feature of the SSR will be called the *input flow control function*.

3.3   The raw materials and parts catalog

The next point to be considered is how the SSR will know which are good raw materials and parts and which are bad materials and parts. The SSR must possess a catalog of good raw materials and parts that will be the informational basis on which the input gateways will open or stay closed. This catalog will be referred to as the *raw materials and parts catalog*.

3.4   The materials and parts identification function

The next issue is how the SSR will recognize and accurately identify a material or part at an input gateway as good or bad. That is not a trivial ability. The SSR needs a way to determine the nature of the materials and parts to which its input gateways are exposed. This ability may be supported by a set of material probing procedures and processes. This SSR ability will be called the *materials and parts identification function*. The complexity of this capability can be compared to the probes on the Martian rover that were used to analyze soil samples for particular compounds.

3.5   The systematic labeling/tagging of all raw materials, raw parts, and fabricated parts and components

An input gateway, assisted by the raw materials and parts identification function, determines that a piece of raw material is one of the good materials recorded in the good raw materials and parts catalog. This piece is going to be admitted into the SSR and transported to a particular place for processing or possibly to a temporary storage location followed by processing. In order to do this successfully, the SSR needs to tag or label this piece so that its nature, once determined at the input gateway, is available for subsequent processing stations or storage stations in the SSR. Therefore, any raw material or part that is allowed to enter the SSR, once its nature is identified, is immediately tagged or labeled using a system similar to the bar codes or RFIDs (radio-frequency identification) where the code used is one of the codes in the catalog of raw materials and parts. This systematic labeling and tagging of all accepted materials and parts will be considered another responsibility of the *materials and parts identification function*. Additionally, during the SSR growth and cloning phases, the SSR will fabricate new parts, components, and assemblies using either raw materials and raw parts or previously fabricated parts, components, and assemblies. The point is that the *materials and parts identification function* will be responsible for tagging or labeling not only raw materials and parts accepted inside SSR, but also all fabricated parts, components, and assemblies. Therefore, all elements inside the SSR and all SSR parts should bear a permanent identification tag.

3.6 The catalog of fabricated parts, components and assemblies

This raises another important aspect for the SSR design: The SSR must possess not only an exhaustive catalog of all raw materials and raw parts, but also a *catalog of all fabricated parts, components and assemblies* with a unique identifier for each *type* of such elements.

3.7 The bill of materials function

The automated fabrication processes will need additional informational support. This capability can be called the *bill of materials function*. It is supported by an exhaustive catalog with entries that specify the list of materials required to fabricate each fabricated part, component,and assembly. For each item in this list, the quantity of those materials must be also specified. The *bill of materials* is an *informational function* of the SSR. Like all SSR information functions, it has two components: a specific catalog (or database), and a set of information access sub-functions to search, read, write, update, or delete specific entries in the associated catalog that can be accessed by all other SSR functions.

In this case, the catalog is the bill of materials catalog. Below is an example of what an abbreviated entry for a "power supply enclosure" part may look like in such a catalog:

Table 9.1: Example Entries in the Bill of Materials Catalog for the "Power Supply Enclosure" part

Part name	Part ID	Flags	Count	Qty	Dimensions
Sheet metal $\frac{1}{16}$	ID-02409	$Q + D$	-	1.2	16x24
Screws $\frac{1}{8} * 2$	ID-01670	C	8	-	-
Washers $\frac{1}{4} * 2$	ID-05629	C	8	-	-

The *flags* field tells what properties are specified for the part.
Q=quantity, C=count, D=dimensions

3.8   The fabrication material extraction function

Some of the raw materials admitted inside the SSR cannot be used directly by the SSR fabrication processes. They need to be transformed into fabrication materials through one or more specific processes. An example, which is not necessarily related to our artificial SSR, is the fabrication of steel (fabrication material) from iron ore and coal (raw materials). The SSR's ability to extract fabrication materials from sets of raw materials and parts is the *fabrication material extraction function*. Fabrication materials are registered in the *catalog of fabrication materials* while every process that is used to extract fabrication materials from raw materials and parts is documented in a *fabrication material extraction process catalog*.

3.9   The supply chain function

One can now consider the case when during the cloning phase the SSR must fabricate a component of type A. The bill of materials entry for a component of type A specifies that its "fabrication recipe" requires 2 raw parts of type X and 4 raw parts of type Y. The SSR will need to be able to coordinate the input gateways to admit prior to the cloning process required quantities of parts X and Y, creating stock in a SSR stock room so that the fabrication of component A can go smoothly. This will allow the fabrication of the component to depend less on the available raw parts at any given moment at the enclosure input gateways. This SSR ability will be referred to as the *supply chain function*. This function is responsible for interacting with fabrication processes within the SSR to gather information prior to fabrication regarding what raw materials, raw parts, or fabricated parts are needed. It will then command other functions (i.e., the input flow control function and the material and parts identification function) to admit, supply, and stock those elements within the SSR.

3.10   The energy generation and distribution function

All the machines inside the SSR need an energy source to perform their work. Thus, the SSR must have the ability to produce energy from the appropriate raw materials and raw parts. This ability will be named the *energy generation and distribution function* since it has the responsibility not only to generate energy, but also to manage and distribute energy to the energy consumers within the SSR. Importantly, the catalog of raw materials and raw parts, as well as the catalog of fabricated parts, may contain entries that are marked as elements used for energy generation and/or distribution. Also the *catalog of processes* will contain entries detailing the material processes that are used to generate energy from the energy-marked materials and parts. The supply chain function is responsible for managing the timely supply of materials and parts not only for fabrication but also for energy generation and transport.

3.11   The transport function

The functioning SSR features multiple sites where specific actions happen. These sites will be distributed spatially inside the SSR or on its enclosure and will have well-established positions relative to the elements that maintain the three dimensional structure of the SSR as it grows (named *scaffolding elements*). For example, there are input gateway sites, possibly some stock room sites, fabrication sites, and assembly sites where the elements of the growing clone inside the SSR are being put together by some machinery. The SSR must have the ability to carry various elements between sites. This ability is named the *transport function*. It may employ specific means of transport, such as conduits, avenues, conveyors, etc., that are adequate for the particular elements that are being transported and the particular sites within the SSR. An important aspect of SSR activity is the transport of information between producers and consumers within the SSR. For this reason, the transport function is also responsible for *transporting information* between the SSR sites, at least for the provision of the physical, lower layers of the transport of information.

3.12   The manipulation function

Another important ability that the SSR needs is the *manipulation function*. This function consists of the ability to handle, grab, or manipulate raw materials, raw parts, fabricated parts, fabricated components, and fabricated assemblies. For example, this ability is needed to take a raw material or raw part admitted at an input gateway and place it on a conveyor that goes to a stock room or a fabrication site. Once there, another manipulator will grab the material or part and place it in a specified position in the stock room or place it on a fabrication bench or machinery. Manipulation examples abound, since no matter what elements are processed, transported, fabricated, assembled or pushed out of an output gateway, there is a need to adequately handle those elements.

3.13   The fabrication function

Since the SSR must be able to clone its core elements, its enclosure, and its scaffolding elements, it is absolutely necessary for the SSR to have a *fabrication function*. This function is the ability to fabricate exact copies of all parts, components, and assemblies that exist inside a mature SSR or on its enclosure. In other words the fabrication function must be able to fabricate all machinery inside the SSR, including fabrication machinery. Since all the elements that reside inside the SSR must be copied (cloned) and various types of information and software elements also reside inside the mature SSR, the fabrication function must also have the ability to accurately copy information and software.

3.14   The assemblage function

Another capability that must reside within the SSR is the *assemblage function*. This function allows the SSR to assemble parts into components and assemblies that increase in complexity. The assemblage function is strongly related to the fabrication function. While these two functions can be seen as two sides of the same coin, it makes sense to see them as distinct functions where the fabrication function creates new parts from raw materials and raw parts through special processes (e.g., metal machining) while the assemblage function puts fabricated parts together into more and more complex assemblies. The assemblage function may be needed, for example, to erect and expand the scaffolding and the enclosure during SSR growth along with creating assemblies of smaller components and parts.

4   Additional SSR Functions

4.1   The recycling function

The SSR functions discussed so far provide specific assistance for the ingestion of new raw materials and parts into the SSR and SSR growth during the cloning phase based on continuous production of energy and planned fabrication of the elements of the daughter clone growing inside the expanding SSR enclosure. As in any process that performs fabrication and construction of new parts, there will be residual raw materials and raw part fragments. The SSR must be designed to carefully control the growth of the internal and enclosure elements. It cannot grow without limits or in an uncontrolled manner. In order to achieve this objective, the SSR must:

- Re-introduce in the fabrication and growth cycles certain elements of the residual raw materials, raw parts, or raw part fragments that can be reused.

- Identify and specifically mark as refuse certain residual elements that cannot be recycled and then expel them as refuse through the output gateways.

- Provide specific processes to clean the SSR interior fabrication and transport spaces such that fabrication and assembly of the cloned parts is not affected and the SSR maintains the proper structure during the cloning and division phases.

Most of the above responsibilities pertain to the *recycling function*.

4.2   The output flow control function

The recycling function controls the *output flow control function*, which is the ability to control the enclosure *output gateways* that force out of the SSR the raw materials, raw parts, and raw part fragments *marked as refuse* by the recycling function.

4.3   The construction plan function

It has already been noted that the bill of materials function provides for each fabricated part, component, or assembly of the SSR a list of raw materials, parts, and sub-components that are needed for fabrication of that element. Thus, an entry in the bill of materials information catalog is similar in concept to the list of ingredients for cooking a meal. Besides the list of ingredients, the recipe for a meal contains an ordered list of steps needed to prepare the meal. In a similar manner, the SSR must store descriptive information for all construction steps needed to fabricate each SSR element. The SSR's ability to store and make accessible detailed information about the set of fabrication steps and processes needed for the fabrication of each SSR element (part, component, assembly of components, including the SSR itself) is called the *construction plan function*.

4.4   The construction plan information catalog

The *construction plan function* has an associated *construction plan information catalog*. This catalog has an entry for each fabricated element of the SSR. A fabricated element can be a simple fabricated part (i.e., fabricated from a single good material). A fabricated element can also be a component which, in this context, refers to an element fabricated through the assemblage of two or more fabricated parts. A fabricated assembly is even more complex: it is fabricated from multiple simple parts and one or more components and possibly one or more (sub) assemblies. The mature SSR (before starting the cloning process) is a particular case of a fabricated assembly. Another example of a fabricated assembly, and the most complex example in this case, is the SSR that contains both the mother core elements and the daughter elements just before division begins.

Each entry in the construction plan catalog contains a reference to the entry in the bill of materials catalog for the same element (to access the parts and components needed for the element fabrication) and an ordered sequence of fabrication and assembly steps.

For each fabrication/assembly step the following information may be provided:

- The spatial assembly or placement instruction of parts and/or materials involved in the step (i.e., how to spatially place a part/component, P, relative to the assembly under construction on the work bench before a fabrication step)

- The type and detailed description of each fabrication/assembly process executed during the current fabrication step

- Technological parameters of the fabrication/assembly process (e.g., ambient temperature, length of process, etc.)

- List of residual parts/materials resulting from the process or particular step

- Part manipulation steps (This includes x, y, z starting point, x, y, z ending point, part rotation, axis specification, or translation movements.)

- Any fabrication step verification procedure to determine if the step completed successfully within accepted parameters or if the fabrication step was a failure.

- Recovery action list in case a fabrication step fails with a specific verification error

In short, the rationale for the nature, structure, and extent of information items stored for each step of a *fabrication plan entry* is to provide support for full automation of that element of fabrication. And, as suggested above, *fabrication steps* can be of a very large variety. The nature of the fabrication process, as well as the nature of fabrication steps, depends on the material basis that will be selected for the design and implementation of the artificial SSR. The alternative material bases that can be realistically considered for creating an artificial SSR are discussed in Part 2 of this study.

Here are several examples of the nature of a fabrication step:

- A positioning step (i.e., a part or component being placed in a particular position relative to the semi-assembled element in preparation of another step)

- An assemblage step

- A metal machining step

- A chemical process step

- A thermal process step

- An electrolytic process step

- A nanotechnology assemblage step

- An information file copy step

- A fabricated component (assembly) test step

The next issue to solve requires devising a way to track the progress of the cloning and division steps. This is provided by the *construction status function*. This function uses an information catalog that is similar to the construction plan catalog named the *construction status catalog*. It has the same list of element entries as the construction plan catalog and describes the same hierarchical composition of each element (part, component, assembly) in sub-elements. Each entry in this catalog has construction status information that reflects the current construction status of that entry and can have values like "not-started," "started," or "completed."

In a similar manner each fabrication step of an entry has a current construction status field that is also used to mark and keep track of the fabrication/construction status for that element at the fabrication step level.

4.5   The SSR variable geometry

Another issue in developing the artificial SSR is that the SSR has variable geometry. The SSR has variable geometry because, first, the mature SSR must grow in volume and enclosure surface during the cloning phase to make space in its interior for the growing clone. Second, the geometry changes even more radically when the division phase starts and culminates with the complete division of the original SSR in both the mother SSR and the daughter SSR.

The SSR variable geometry presents many considerations and challenges that must be carefully considered by the SSR design and by the hypothetical implementation of the artificial SSR. The first consideration is the structure and composition (texture) of the SSR. This must be designed to allow the following:

- Surface area growth as the SSR interior grows in volume during the cloning phase. This may require a design that allows selective insertion of new enclosure parts/elements in between existing parts/elements and a way to link or connect each new element with its neighboring elements.

- Insertion of new input gateways and output gateways while the enclosure grows.

- Division of the enclosure into two separate enclosures (one for the mother SSR and one for the daughter SSR) with each enclosure carrying its separate sets of input and output gateways, scaffolding, and interior elements.

Next, special design provisions must be made for the interior SSR scaffolding. The SSR scaffolding is made of structural elements (e.g., pylons, walls, supports, connectors, etc.) that are needed to maintain the three-dimensional structure and integrity of the enclosure and of the SSR interior space(s). The scaffolding design and its elements need to be conceived such that, first, the scaffolding elements may change size as the interior of the SSR grows (during cloning) or shrinks (during division); second, the spacing between scaffolding elements and their connectors may also grow or shrink (during cloning or division phases). Finally, the design of the SSR must also make provisions for the growth, variable geometry, and dynamic restructuring and re-linking of any SSR transport, conduits, paths or communication lines during the cloning and division phases.

The variable geometry means that the SSR design must make specific, detailed provisions for the entirety of its spatial evolution. This includes all geometrical definition points or trajectories (in the x, y, and z axes) of all variable elements of the SSR enclosure, scaffolding and interior. These spatial trajectories need to be harmoniously and coherently coordinated with all fabrication and assemblage steps of the cloning and division phases.

4.6 The Communication and Notification Function

This function is responsible for providing and managing the information communication and notification machinery and mechanisms between the command centers and execution centers of the SSR. For example, the fabrication control function, acting as a command center, may send a command as a specifically encoded information "package" to the fabrication and assemblage functions to build a particular component of the daughter clone. In this circumstance, the fabrication and the assemblage functions operate as execution centers for the command. When the fabrication of the requested component is completed by the fabrication function, it will send a specifically formatted notification information package back to the fabrication control function indicating that the specific command for the fabrication of the specific element was successfully completed. Within the same example scenario, the fabrication function will send, in its turn,—this time playing a command role itself—a command to the supply-chain function (i.e., the executor entity) to trigger the transport of the needed fabrication ingredients for the clone element to the fabrication site. The *communication and notification function* will need to be deployed ubiquitously throughout the SSR to allow communications/notifications between various functions and machinery operating all over the SSR.

5 The Higher Level SSR functions

The SSR functions that are described in subsequent sections are the highest level functions of the SSR. They accomplish their goals by coordinating and choreographing the lower level functions described in the previous sections.

5.1 The scaffolding growth function

The *scaffolding growth function* is responsible for managing the construction, growth, and position change of the SSR scaffolding elements during the cloning and the division phases of the SSR replication process. This function needs to manage the variable geometry of the scaffolding elements in synchronization and coordination with the other spatial changes of the SSR both on its enclosure and within its interior.

5.2 The enclosure growth function

The *enclosure growth function* is responsible for managing the construction, growth, and shape change of the SSR enclosure as well as the coordinated addition of input and output gateways on the enclosure during the cloning and division phases of the SSR replication process. As mentioned before, this function needs to manage the variable geometry of the enclosure, the dynamic shifting of the gateways on the enclosure's surface, and the enclosure's radical shape changes during the division of the SSR into the mother and daughter descendants.

5.3   The fabrication control function

The *fabrication control function* is responsible for the construction, assemblage and variable geometry management of all interior elements, in particular those related to the cloning portion of the SSR involved in the cloning and the division phases. Like the two preceding growth functions, this function coordinates activities of the fabrication function, assemblage function, construction plan function, recycling function, and other lower level functions.

5.4   The cloning control function

The *cloning control function* is responsible for coordinating the whole cloning phase of the growing SSR. It coordinates the cloning and the growth of all of the involved SSR compartments through tight control, synchronization, and coordination of the scaffolding growth, enclosure growth, and fabrication control functions. In particular, this function is responsible for starting the cloning process, monitoring its development, and accurately determining when the cloning process is complete. One particular responsibility of the cloning control function is the *cloning of the information* stored in the mother SSR. This information cloning is performed at the end of the cloning phase when all internal machinery, internal scaffolding and enclosure elements of the mature SSR are completely cloned and constructed as part of the nascent daughter SSR. The information is cloned by systematically, accurately, and completely copying all information catalogs from resident machinery of the mature SSR into the corresponding new machinery of the clone part.

5.5   The division control function

The *division control function* has full control of the division phase of the SSR replication process. It manages the SSR division through specific commands sent to the scaffolding growth, enclosure growth and fabrication control functions. In particular, this function is responsible for starting the division process, for choreographing its development on all SSR compartments (enclosure, scaffolding and core), and accurately determining when the division process is complete. One particular responsibility of the division control function is to "start the engines" of the nascent daughter SSR. Just before the moment the division is complete, the division control function must send a command to the daughter SSR to start its own machinery and control functions. When the division completes, the separated daughter SSR becomes a mature, fully functioning autonomous SSR ready to start its own replication process. The separated "mother SSR" can start, in its turn, a new replication cycle.

5.6   The replication control function

The *replication control function* is the highest level SSR function. It is responsible for accomplishing the full SSR replication cycle by coordinating and choreographing its two phases: cloning and division. It does this through corresponding control and coordination of the *cloning control function* and the *division control function*.

Metaphorically speaking the replication control function implements the two significant *SSR designer commandments*: grow and multiply.

5.7   The SSR Function Dependencies Diagram

Figure 9.4 provides an overall diagram illustrating the identified set of functions present in the SSR and some of the dependencies between these functions. This figure depicts the main dependencies and interactions between the identified SSR functions, but is limited to only the primary dependencies and interactions.

The communication and notification function is represented separately since it relates to and is used by almost all other SSR functions. Its ubiquity throughout the SSR functions is due to the need for information communication between functions as well as notification (another form of information exchange).

As already mentioned, the relationships and dependencies between the functions are more complex than depicted in the diagram. For example, the transport function depends on the energy generation and transport function although this dependency is not depicted in the diagram.

6   The Type and Nature of SSR Components

At this point it makes sense to break down the previous discussions into the general conceptual categories of the artificial SSR's components to provide an overview of what needs to be accomplished for its creation. Part 2 of this paper will go into more concrete details about the physical components that may make up the artificial SSR. The categories themselves are grouped by functional area.

6.1   Information Processing Component Categories

- Information storage and access

- Information processing

- Information coding and decoding

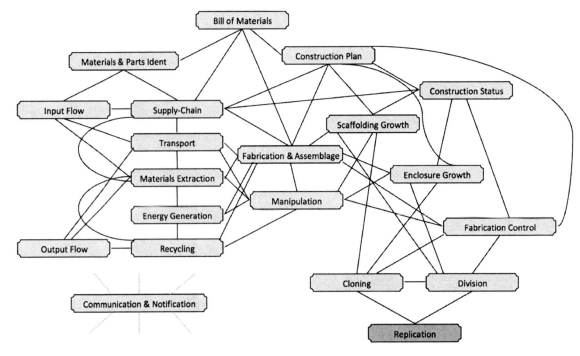

Figure 9.4: SSR Functions and their Dependencies

- Information transport, communication and notification

6.2   Materials Component Categories

- Material identification

- Material transport and manipulation

- Material processing

- Mechanical and chemical transformation of materials

- Material fabrication and assemblage

6.3   Energy Component Categories

- Energy generation

- Energy transport

- Energy conversion

- Energy distribution and management

6.4   Environmental Component Categories

- Environment sensing

- Environment (local) control

6.5   Construction Component Categories

- SSR construction plan representation

- SSR dynamic 3D evolution representation

- SSR construction status representation

- SSR parts inventory representation

7   The SSR and Its Information Catalogs

Various types of information catalogs (databases or repositories) that together make up the information base of the SSR have already been presented. This section contains a list of all of the types of information catalogs identified thus far. It is quite probable that after a more in-depth analysis of the topic, there may be the need for additional types of information catalogs.

The SSR information catalogs need to thoroughly, systematically, and coherently capture all information describing in detail each element of the SSR during its full life-cycle, all relationships between these elements captured in construction plans (body plans), and all fabrication and assemblage procedures and processes.

This requires making certain simplifying assumptions in order to make the presentation of ideas easier to understand. This means that every piece of data or information that the SSR design must capture cannot be fully dissected here. A real information repository not only contains conceptual lists of tables (catalogs) of items of the same nature but also lists of the various relationships that exist between the items in different tables (catalogs). For example, there are certain relationships between the items in the catalog of raw materials and the items in the catalog of the bill of materials. There are different sets of relationships between the entries in the catalog of the construction plans and the entries in the catalog of the bill of materials and the entries in the catalog of processes. For now, this is just a list of the core catalogs that have been identified so far:

- The catalog of raw materials

- The catalog of raw parts

- The catalog of fabrication materials

- The catalog of raw materials identification procedures and processes

- The catalog of raw parts identification procedures and processes

- The catalog of fabrication materials extraction procedures and processes

- The catalog of energy generation procedures and processes

- The bill of materials catalog

- The catalog of internal machinery

- The catalog of construction plans

- The catalog of fabrication procedures and processes

- The catalog of assembly procedures and processes

- The catalog of fabrication verification procedures

- The catalog of fabrication errors handling procedures

- The catalog of energy consumptions

- The catalog of recycling elements and procedures

- The catalog of construction statuses

8   Conclusion

This part provides an abstract description of the minimal core components, processes, information stores, and structural requirements of an artificial self-replicating system. Part 2 will cover physical considerations for implementing such a design, and Part 3 will cover speculative ideas for what the existence of self-replicative processes in nature indicates on the larger scale.

10

Developing Insights into the Design of the Simplest Self-Replicator and Its Complexity: Part 2—Evaluating the Complexity of a Concrete Implementation of an Artificial SSR

ARMINIUS MIGNEA

The Lone Pine Software

Abstract

This is the second in a three-part series investigating the internals of the simplest possible self-replicator (SSR). It builds on the construction of a hypothetical self-replicator devised in Part 1 and considers various significant aspects about the design and construction of an artificial, concrete SSR: the material basis of its construction, the effects of the variable geometry of the SSR during its growth through the cloning and division phases, and the three closure rules that must be satisfied by the SSR—energy closure, material closure, and information closure.

The highest technical challenges that need to be faced by the design and construction of the artificial SSR are considered. The emerging complexity of the artificial SSR is depicted using a metaphorical comparison of the SSR with a city fully populated by automated machinery that systematically constructs a new city that is identical to the old city without external help but only using the construction materials that enter through the city gateways. The current level of technology is evaluated to determine if it is sufficient for the successful completion of the design and construction of an artificial autonomous SSR project using either a nano-biochemical basis or or a macro-material basis.

Part 1 of this series analyzed the basic necessary design elements of the simplest self-replicator (SSR), including necessary components, functions, processes, and information. Having established the minimum requirements for the design, this part will discuss the physical implementation of the SSR.

1 The Three Closure Requirements as the Basis of an Autonomous SSR

The SSR must be fully autonomous. This means that it can only obtain raw materials and raw parts from its environment and benefit from (or struggle because of) the environmental conditions specific to its location.

Full autonomy specifically requires the SSR to exhibit the following characteristics (Freitas Jr. & Merkle, 2004):

1. The SSR must fabricate all its energy from input materials, and the generated energy must be sufficient for the SSR to produce an exact replica of itself. This condition is called the *energy closure*.

2. The SSR must use only materials admitted through its input gateways, and these materials must be sufficient for the SSR to grow and generate its daughter. This condition is called the *material closure*.

3. The SSR must use only information that is initially present or stored in the mature SSR, and this information must be sufficient to produce an exact replica of the SSR. This condition is called the *information closure*.

2 The Core Approach to Cloning

This section will try to answer the following important question for the design of the artificial SSR: What is the core mechanism that the artificial SSR will use to accurately clone all of its elements?

Below are two possible answers, and it is likely that most any other imagined answers would be similar or equivalent to one of the two answers below:

1. Design and use a *universal physical copy machine* (similar to a key copy machine but much more sophisticated) that analyzes each part or assembly and produces a copy, with the goal of simplifying the design and avoiding having to maintain such detailed catalogs of information.

2. Use an exhaustive descriptive, operational and constructional SSR information database directing an integrated set of specialized software and computer-controlled automatons, in other words, have sufficient data stored about the makeup of the SSR itself to generate a new copy from that data.

2.1 Why the "universal physical copy machine" approach is not adequate

This approach assumes that the SSR contains a sophisticated machine that can examine and accurately copy all other pieces and machinery comprising the mature SSR. This implies that this universal physical copy machine can even copy itself or, more realistically, make a copy of a copy of itself. In other words, the SSR contains two universal physical copy machines. One, Machine A, is performing the actual copying of all SSR machinery and the other universal physical copy machine, Machine B. So the second copy machine, Machine B, is only used as a model for the first Machine A. The goal for this solution is to use these copy machines A and B to alleviate the need to store as much information about the SSR itself, to have as much software, and to contain as much computer-controlled machinery as in the second approach.

This first solution leads to the following conclusions:

1. It will require another machinery $M^{disassembler}$, to disassemble Machine B in all its constituent parts: b_1, b_2, b_3, . . . , b_N so that Machine A can copy each constituent part.

2. Then Machine A will copy all parts b_1, b_2, b_3, . . . , b_N twice (to create all pieces needed for a clone of Machine A: $Copy_A$ and a clone of Machine B: $Copy_B$) ca_1, ca_2, ca_3, . . . , ca_N (for $Copy_A$ machine) and cb_1, cb_2, cb_3, . . . , cb_N (for $Copy_B$ machine).

3. There will be a need for another machinery $M^{assembler}$ that will know how to take all the parts ca_1, ca_2, ca_3, . . . , ca_N and assemble them together into the $Copy_A$ machine and parts cb_1, cb_2, cb_3, . . . , cb_N and assemble them together into the $Copy_B$ machine.

In order to properly construct $M^{assembler}$, a vast amount of well-structured information of this nature must be first programmed:

- A catalog of all parts b_1, b_2, b_3, . . . , b_N and for each such part, a unique identifier and possibly physical and geometrical characteristics (dimensions) of the part

- A store room location (x, y, z) from where the $M^{assembler}$ machine will pick each one of the parts ca_1, ca_2, ca_3, . . . , ca_N during the assembly steps to construct the $Copy_A$ machine

- A *catalog of assembly instructions* that contains some geometrical x, y, z instructions and the type of assemblage step (e.g., screwing, inserting, welding, etc.). For example, these assembly instructions may describe processes such as the following:

    - how to put together part ca_2 to the assembly made of parts: (ca_1);
    - how to add part ca_3 to the assembly made of parts (ca_1, ca_2);
    - how to add part ca_4 to assembly made of parts (ca_1, ca_2, ca_3).

  Such descriptions would continue for all parts that need to be included in the assembly.

- A manipulator machine (robot that can follow computerized instructions) to be controlled by the $M^{assembler}$ machine in assembling the $Copy_A$ machine

Although the original impetus for Solution 1 is to avoid mountains of information and armies of automatons and machinery, this solution requires those ingredients. There is no magic copy machine that can do its work without structured collections of information and many helper automatons (machinery) that, in turn, must be information-, software-, and computer-controlled. The conclusion is that there is no "magic universal physical copy machine" solution that is significantly distinguishable from the Solution 2.

Another problem with Solution 1 is that certain components of Machine B may not be fabricated by plain (mechanical) assemblage of parts but rather by using more demanding assemblage processes, such as welding or electro-chemical processes. These processes have no precise means of disassembly, which would prevent the universal physical copy machine from being able to reproduce them.

2.2 Exhaustive information, integrated systems driving information-controlled automatons

Since the universal physical copy machine approach does not work, the only other replication method designs the SSR as a collection of integrated sub-systems, controlling a large variety of automatons using a significant collection of integrated information catalogs (databases).

A number of information catalogs have already been mentioned while identifying specific SSR functions. Any informational SSR function has both an associated catalog and also a set of access sub-functions that provide a set of access operations to the information catalog that can be used by other SSR functions to execute specific action sequences.

3   The Material Basis of the SSR

When approaching the task of the design and implementation of an artificial SSR, a capital question surfaces rather quickly: What should be the material basis for the artificial SSR? There are two distinct possibilities: Either use a biological basis for the SSR on a micro/nano scale or use more common macro scale materials and technology.

3.1   Using a Biological Basis for the SSR

Using a biological basis for the SSR means that the SSR must be constructed using organic materials. These materials would be the same or similar to those used by the cells, tissues, and organs of the living world. There are several advantages to this approach. First, the proof in the feasibility of this approach is the presence of the varied organisms and microorganisms in nature. A key question is whether this approach is accessible given current engineering technologies. Second, even though energy generation is one of the biggest challenges for SSR replication, there are known levels of energy consumption based on biological systems. Lastly, biological systems tend to be in aqueous medium, which may facilitate solutions for the variable geometry problem.

The biological approach is not without its setbacks. First, since this approach operates at extremely small scales, it taxes the limits of current investigative tools and observational methods. The most advanced microbiology manipulation and fabrication tools/approaches are still rather primitive and very limited when considering the tasks that need to be accomplished: fabrication, assemblage, manipulation at nano scales, computing machinery fabrication, software execution, information storage, and communication. Secondly, there are many aspects of cell biology that are still beyond current understandings of the cell's function. Examples of the challenges that we cannot solve with current technology include

- building computers on a biological material basis/scale (or understanding how the cell proteins and other organic cell elements can be used for computation in a general way)

- building bio-chemical manufacturing machines at a biological scale

- building information storage using biochemical materials

- having software running on biological type computers

- communicating information on a biological material basis and at a biological scale

Clearly, the conclusion is that, with the current level of technology, it is impossible to create a design and an implementation plan for an artificial SSR using a biochemical and a biological material basis at a molecular scale. Other alternatives must be considered for a better chance of successfully building an artificial SSR.

3.2   Using a Macro Scale Basis for the SSR

The other alternative for consideration is the macro scale approach, using materials and technology that are in common use for product fabrication. This approach must consider the minimum dimensional scales for which there are available manufacturing technologies for most of the parts, components, and machinery that make up the SSR.

The materials used to construct the SSR enclosure, SSR scaffolding, and SSR interior should be common engineering materials used by current fabrication technologies: metals, alloys, plastics, ceramics, silicon or other special materials. The scale of these artifacts to be fabricated as elements of the artificial SSR must be selected with care as there are two opposing considerations which must be balanced and compromised.

The first consideration is that the smallest possible scale should be used in the design and implementation of the artificial SSR parts in order to minimize the energy consumed by the SSR during a replication cycle and minimize the size, volume and mass of the artificial SSR in order to minimize the number of materials ingested into the artificial SSR and used for fabrication of the clone within the SSR.

However, this must be balanced with the limitations of known engineering technologies and machinery to fabricate, manipulate and assemble all the parts of the SSR machinery. This means, for an illustrative example, that if the minimum size of semiconductor fabrication equipment that is being manufactured today is 0.5 meters, then the designed size of the mature artificial SSR cannot be smaller than 1 meter. Therefore, a more realistic artificial SSR design would have dimensions in the range of at least 10–100 meters.

4   The Type and Nature of SSR Components

The conclusion in Part 1 was that the SSR must be designed and implemented as a collection of integrated, computer-controlled and software-controlled automatons. The artificial SSR must, by necessity, contain these types of elements:

- Computing machinery, which implies that the following type of elements must be present inside the artificial SSR:

  - Printed circuit boards (PCB)
  - Microprocessors

- Highly integrated circuits (Application Specific Integrated Circuits = ASICs)—specialized, high density integrated circuits for specific computing/application tasks, such as networking, numerical processing, image processing, etc.
- Semiconductor memories (solid state memories)
- Magnetic memory (hard drives)
- Electric power supplies
- Computer connectors and wiring

- Networking Communication Devices:

  - Routers (wired/wireless)
  - Switches
  - Modems

- Software

- Robots

- Energy generation and distribution machinery

  - Generators
  - Transformers
  - Converters
  - Wiring

- Batteries

- Fabrication machinery

- Metal machining machinery

5   Derived Design Requirements

This section is a list of design and implementation requirements for the artificial SSR that emerged from the previous analysis and from the inferences presented so far. These requirements were only implied during the discussion so far but are now made explicit and are described in some detail.

5.1   Each SSR Machine is Power-Driven

If each SSR machine is power-driven, then certain significant consequences arise in designing an artificial SSR:

- The SSR must have a power distribution network (e.g., an electrical distribution network) that must reach each of the SSR's machinery. The design of the layout and geometry of the power network must consider the variable geometry of the SSR enclosure, scaffolding, and interior space and structure. Particular consideration must be made for the zones affected by growth and shape changes.

- Each SSR machine must be designed to use and consume power (electricity) at a level adequate for its nature and the actions it performs.

- The machinery is computer driven, which means that either there is a parallel SSR power network for an energy level (e.g., voltage) adequate for computing devices, or each machine must have some adequate power converters (e.g., electrical power supplies or batteries).

- The SSR machines that provide mechanical work or movement must be provided with motors (rotational and/or linear) adequate for their nature.

- Mobile machinery (e.g., transporters, moving robots) must be designed such that their mobility is not constrained while they are connected to the SSR power network(s). Designing all mobile machinery with rechargeable batteries may solve or significantly simplify the connectivity constraints but will require additional provisions for battery fabrication processes and fabrication and provision of battery charging stations.

- The design of each machine must provide specification for the average power consumption on all power networks (normal power level and computer power level) to which the machine is connected.

5.2   Each SSR Machine is Computer-Driven and Software-Driven

If each SSR machine is computer-driven and software-driven, then there are several consequences to consider in designing an artificial SSR:

- Most SSR machines must host at least one internal computer with the possible exception of some simpler (e.g., electro-mechanical) machines that can be remotely controlled.

- The SSR must have highly technical machinery and processes to fabricate computers and their respective parts.

- Each SSR machine that hosts computing devices must be networked—by wire or wirelessly—to other machines and control centers within the SSR.

- The SSR must have adequate machinery to not only fabricate computers but to install them into other SSR machines, plug them into the other machine's power network, and connect them to the SSR's communication network.

- The SSR must have machinery that is able to download and copy software into any computer installed into an SSR machine, to start (boot) that software on the machine, and to monitor its availability and behavior.

- The SSR must have the capability to test each piece of its machinery, to detect malfunctions in computer and software installations as well as in the machine hardware, to detect malfunctions in the computer and software execution and to have adequate procedures to diagnose and repair the identified problems. Diagnosis and repair may be based on the availability of fabricated spare parts.

- The software that drives each particular piece of machinery must be designed and written with a full understanding of the physical and cinematic capabilities and constraints of that piece of machinery. It must take into account all possible uses of the machine and all its components' behaviors and interactions with external objects and events as well as be able to handle them correctly.

- The software that drives fabrication and assemblage machinery and materials and fabrication processes needs to be based on a thorough design of the machines that will be built, their cinematic capabilities, and their specified power and energy consumption.

- The software written for various SSR functions must carefully and accurately coordinate and synchronize the activities of multiple SSR machines (e.g., fabrication machines, material process machines, manipulation and transport robots/arms, assemblage and construction machines) by providing a continuous monitoring of the 3D spaces occupied by each machine and its mobile parts to avoid collisions and to ensure cooperative progress with both lower level and higher level tasks of the growing SSR.

5.3 Each Piece of SSR Machinery Is Capable of Information Communication

If each piece of the SSR's machinery must have the ability to communicate information, then several features must be included in its design:

- The artificial SSR is a collection of automated machines and robots. Their cooperation and coordination for achieving tasks from the simplest (e.g., fabricating a part, or manipulating a part in a sequence of steps for an assemblage operation) to the most complex ones (e.g., the fabrication and assemblage of computing hardware and software installation for a new piece of fabrication machinery) requires extensive, continuous, multipoint, and multi-level communication of information between machines, control functions, and software components.

- The SSR must have a comprehensive physical layer communication network for information transport (wire-based and/or wireless) with access points located on each, if not most, SSR machines/robots and sometimes in between the subsystems of the same SSR machine.

- The SSR might need to have adequate networking devices (e.g., routers, switches, modems, and codecs) to implement needed communication patterns and topologies.

- The SSR machine and software components engaged in communication will need adequate networking/communication protocols with appropriate characteristics for carrying the needed communication bandwidth, handling errors and retransmissions, reliable routing, and end point addressing.

- The SSR should have the ability to deploy software on newly constructed machines and network nodes, and to bring up the network and verify it as part of starting up the daughter SSR system (including its underlying communication network) as a preparatory step in the SSR division phase.

6 The Most Significant Challenges for the Design and Implementation of an Artificial SSR

6.1 The energy generation and the energy closure challenge

The energy generation and energy closure challenge presents multiple hurdles which must be resolved. Primarily, the SSR design must select an adequate basis for energy generation. This depends on what natural materials are available in the SSR environment that can be used for energy generation. Some of the candidate material basis for energy generation that might be considered for the design and implementation of an artificial SSR might include biochemical or organic (e.g., vegetation used for energy generation), coal, oil/petroleum, natural gas, methanol, hydrogen, solar, wind, and/or nuclear.

The *energy closure challenge* means that the amount of energy generated by the SSR from the primary energy producing materials extracted from the SSR environment must be sufficient to power all machinery (e.g., fabrication, assemblage, construction, transport, manipulators, robots, computers, and networking gear) that equip the SSR.

Other challenges associated with energy closure include the following:

- The SSR must be designed with the ability to slow down or even completely shut down during the periods when the input of energy producing materials is reduced or null.

- The SSR's ability to provide the means to store energy (with batteries, accumulators or stocking energy-producing materials) may smooth out or eliminate the need for transition to "hibernation" or shutdown states.

- Designing SSR machines with local sources of energy (e.g., rechargeable batteries, accumulators, fuel reservoirs, or fuel cells) may provide true, unconstrained mobility and may significantly simplify the SSR design and implementation difficulties related to keeping all mobile machines hooked to flexible power wiring or network wiring.

- Burning fuels or chemically generating energy leads to additional concerns in designing and implementing an artificial SSR. Particular concerns include preserving the SSR's internal environmental parameters (e.g., temperature and humidity), avoiding hazardous materials, and providing storage transportation containers for liquid or gaseous materials.

6.2   The material closure challenge

The material closure challenge for designing and implementing the artificial SSR can be summarized as follows: all fabrication materials that are needed for fabricating the parts of the SSR machinery must be available in the SSR environment or must be extracted from raw materials available in the SSR environment.

While this may not appear to be a daunting task, upon closer consideration, the artificial SSR will need fabrication machinery for metal machining, computers with semiconductor microprocessors and memories, plastics, and ceramics. There is potentially an extremely long list of materials needed for SSR fabrication. For example, looking at a short subset of materials needed for the macro scale artificial SSR provides insight into the size of the challenge:

- Iron

- Steel (of various varieties)

- Copper

- Aluminum

- Metal alloys (of different varieties)

- Silver

- Gold

- Ceramics

- Plastics

- Silicon

- Polytetrafluoroethylene (Teflon) for Printed Circuit Boards (PCBs)

- Tin

- Nickel

- Germanium

.   Even given the partial list above, it appears that there is a very small probability that the SSR's local environment will feature such a large diversity of immediately available materials or components from which the materials in the list could somehow be extracted. This makes the material closure requirement appear *unsolvable*, and thus any project to design and implement an artificial macro-scale replicator may be condemned to failure.

Another way to formulate the material closure challenge is that a successful selection of materials used for energy generation and fabrication of internal parts, components, and machinery must be based on a thorough knowledge of the environment in which the designed SSR is projected to exist, including the nature of raw materials and parts in such environment, and realistic material extraction paths and processes. In other words, the success requires a perfect design of *both* the SSR and its environment.

6.3   The Fabrication Challenge

The fabrication challenge is the requirement that the artificial SSR must be able to fabricate and assemble any type of parts, components and machines that are part of the mature SSR, which implies that all fabrication and assemblage machines should be able to fabricate exact copies of themselves.

While the material closure challenge focuses on the difficulty of having a wide spectrum of fabrication materials readily available, the fabrication challenge raises several other concerns.

- Since the artificial SSR will have much machinery made with metals (e.g., fabrication machinery, construction and assemblage machines, robots, manipulator arms, networking gear, wires, power supplies, conduits, and scaffolding), the SSR must have a diversity of metal machining machinery.

- The SSR must be able to fabricate the necessary machinery and enclosures for energy generation, as specified above.

- The SSR must be able to fabricate machinery and enclosures to control and host a very wide set of processes (e.g., material extraction, energy generation, possible chemical reaction processes, electrolytic processes, and PCB etching chemical processes).

- The SSR must be able to fabricate computers and computer parts including microprocessors, integrated circuits, application specific integrated circuits (ASICs), signal processing integrated circuits, controller integrated circuits, printed circuit boards (PCBs), power supplies, cabling, semiconductor memories, magnetic memories, and media (hard drives, solid state drives). This also implies that the SSR must feature highly demanding "clean room" spaces that robotically manipulate materials and parts as well as perform semiconductor fabrication.

6.4 The Information Closure Challenge and the Hardware/Software Completeness Challenge

The SSR's information closure is the requirement that the information contained in the SSR is sufficient to drive its successful replication without any additional external information. Completing the hardware and software requirements further extends the information closure requirement by demanding that the computing hardware and software present in the SSR together with the information resident in the SSR are sufficient to drive, control, and successfully complete the cloning and division phases of SSR replication. The SSR's hardware and software must provide full automation of the control, fabrication, assemblage, and the handling of special situations like error detection, error repair, and recovery after error.

The *design of the information* resident in the SSR must be appropriate for its self-replication. Its characteristics must be complete and adequate for the task. It must cover all relevant aspects that intervene during replication (e.g., materials, parts, processes, procedures, plans, spatial structures, error and recovery handling, etc.). Completeness means also that the information designed and stored in the SSR is correctly correlated with the SSR environment. That means, for example, that the SSR design should be based on an accurate and exhaustive list of raw materials and raw parts that exist in the SSR environment together with the material identification procedures and material processing/extraction procedures for those materials. Additionally, it must be adequate for the task in that it must cover all descriptive details of all entries in the information catalogs, with all relevant properties for these entries, including the correct representation of various relationships between the entries in the information catalogs.

The *computing hardware and software completeness requirement* has several implications. The designed computing hardware and software for each machine must be complete, sufficient, and adequate to control, drive, and monitor that particular machine. It must answer commands from the SSR control centers and properly communicate information, status, and control commands with other machines as needed to accomplish the higher level functions of the SSR. Additionally, the hardware and software that are used by various SSR functions and control centers are also complete, sufficient, and adequate in that they cover all possible use cases including errors and incidents.

6.5   The Highest Challenge: The SSR Design Challenge

The SSR design challenge simply means that the SSR's design, including the design of all its subsystems (reviewed in the previous sections), are adequate for accomplishing successful self-replication of the fully autonomous SSR with preservation and without degradation of the self-replication capability that is passed to all generations of daughter SSRs.

There are several specific aspects of the design challenge enumerated below:

- The design of the SSR must be fully coordinated with the design of the environment in which the SSR will be placed. This means in particular that the SSR's design needs to be fully informed about the nature, characteristics, and environmental conditions (e.g., temperature, pressure, humidity, and aggregation status) of the medium where it will exist, including the nature of raw materials and raw parts that are present in this medium.

- The analysis conducted so far reveals that the design and construction of a fully autonomous self-replicating SSR are *extremely demanding*. The success of such a design and construction appear to be heavily determined by the appropriate choices, listed below, and how these choices harmonize with the SSR environment:

  - the material basis of the SSR (nano scale chemical basis or macro material basis)
  - the overall aggregation status of the SSR components: liquid, solid (compact or with embedded spaces), aqueous, colloidal
  - the scale of the mature SSR

- the availability of energy-generation materials and processes in the material basis of choice and at the scale of choice

- the availability of well mastered techniques for the SSR material basis of choice, scale of choice of fundamental engineering techniques including energy-generation and transport, fabrication, assemblage and construction, transport and mobility, manipulation, computation, information communication, and sensing.

7   The Emerging Image of the Artificial SSR

An artificial SSR is very similar to a modern city enclosed in a dome-like structure that communicates with the outside world by well-guarded gates used by robots to bring in construction materials from outside the city. This modern city has two quarters: the "old city" with its fully functional infrastructure in place including buildings, plants, and avenues. The "new city" quarters are initially a small, empty terrain. As the new city is being constructed and its area extends, the dome covering gradually extends to cover both the old, established city and the new, growing city. Both the old city and new city quarters are pulsating with construction activity: automated machines (robots) carry new materials, parts, and components that are used to construct the infrastructure of the new city quarters into an exact replica of the old city and to continuously extend the dome on top of it.

The old city contains the following structures that must be replicated in the new city:

- material mining sub-units

- metallurgic plants

- chemical plants

- power plants

- an electricity distribution network

- a library with information for all city construction plans in electronic form that is made available on the city web and used by the city control centers, its machinery, and robots

- a network of avenues, alleys, and conduits for robotized transportation

- fully automated and robotized manufacturing plants specialized in fabrication of all parts, component assemblies, and machines that are present in the old city

- a fully automated semiconductor manufacturing plant with clean rooms for fabrication of microprocessors, ASICs, memories, and other highly integrated semiconductor circuits and controllers

- a computer manufacturing plant

- a network equipment manufacturing plant

- an extended communication network connecting all plants and robots

- a software manufacturing plant and software distribution and installation of robotized agents

- warehouses and stockrooms to store raw and fabricated materials, parts, components, assemblies, and software on some storage media

- a materials and parts recycling and refuse management plant

- an army of intelligent robots for transportation, manipulation, fabrication, and assemblage

- an army of recycling robots that maintain clean avenues and terrains in the city, collect debris from various plants and reintroduce the recyclable materials and parts into the fabrication process while the unusable parts are taken out of the city gates

- control, command, and monitoring centers that coordinate the supply of materials and the fabrication of an identical copy of the original plants, avenues, factories, and stockrooms

- a highly sophisticated, distributed, multi-layered software system that controls all plants, robots, and communications in a cohesive manner

Each one of the transport carriers, robots, manipulators, and construction machinery is active and does its work without impeding the movement of any other machine. Everything appears to be moving seamlessly and orderly, and the construction of the new city is making visible progress under the growing pylons of the bolting dome.

When the new city quarters are completed and they look like the old city, the machinery and robots of the new city become active, bringing the city to life. Something starts happening as well: machinery and robots from both the old and the new city start remodeling the supporting pylons and the arching dome. What used to be a single, super-arching dome is changing shape into two separate domes one for each of the city quarters.

The final steps require the two teams of robots to coordinate as they complete the separation process of the old city and the new city. The new city has its own dome that has been shaped and separated from the original city. What were once two quarters of the same city is now two completely separate cities starting their own, separate destinies.

8 A Brief Survey of Attempts to Build Artificial Self-Replicators

No successful attempt has been made so far to build a real autonomous artificial SSR from scratch. W. M. Stevens summarizes the situation in the abstract of his PhD thesis:

> Research into autonomous constructing systems capable of constructing duplicates of themselves has focused either on highly abstract logical models, such as cellular automata, or on physical systems that are deliberately simplified so as to make the problem more tractable (Stevens, 2009).

Stevens reviews some of the attempts made at building physical or abstract self-replicating machines:

- **Von Neumann's kinematic model.** The system is comprised of a control unit governing the actions of a constructing unit, capable of producing any automaton according to a description provided to it on a linear tape-like memory structure. The constructing unit picks up the parts it needs from an unlimited pool of parts and assembles them into the desired automaton. The project was far from being finished and remained an abstract model when von Neumann died (Stevens, 2009).

- **Moses' programmable constructor.** Matt Moses developed a physical constructor designed to be capable of constructing a replica of itself under the control of a human operator. The system is made of only eleven different types of tailor-made plastic blocks (Moses, 2001).

- **Self-replicating modular robots.** Zykov, Mytilinaios, Adams, and Lipson built a modular robotic system in which a configuration of four modules can construct a replica configuration when provided with a supply of additional modules in a location known to the robot (Zykov, Mytilinaios, Adams, & Lipson, 2005).

- **The RepRap project and 3D printing**. Bowyer et al. have developed a rapid prototyping system based around a 3D printer that is capable of being programmed to manufacture arbitrary 3D objects. Many of the parts of the printer that are assumed to be self-reproducing cannot be manufactured by the system itself. Those parts happen to be the most complex ones (e.g., the computer controller) (Bowyer, 2007).

- **Drexler's assembler.** In *Engines of Creation*, K. Eric Drexler describes a molecular assembler that is capable of operating at the atomic scale. The molecular machine has a programmable computer, a mobile constructing head and a set of interchangeable reaction tips that will trigger chemical reactions designed to construct any object at molecular scale (Drexler, 1986). This proposal stirred quite a controversy with some scientists who were skeptical of the feasibility of such a project (Smalley, 2001).

- **Craig Venter's synthetic bacterial cell and synthetic biology**. Craig Venter and the scientists at J. Craig Venter Institute in Rockville, MD reported in the May 20, 2010 issue of the journal *Science* that they created a "new species—dubbed Mycoplasma mycoides JCVI-syn1.0—that is similar to one found in nature, except that the chromosome that controls each cell was created from scratch" (Smith, 2010; Gibson et al., 2010). In the same ABC news report, Mark Bedau, professor of Philosophy and Humanities at Reed College in Portland, Oregon, called the new species "a normal bacterium with a prosthetic genome" (Smith, 2010).

- **Synthetic biology** is a new area of biological research and technology that combines science and engineering. It encompasses a variety of different approaches, methodologies, and disciplines with a variety of definitions. The common goal is the design and construction of new biological functions and systems not found in nature (Heinemann & Panke, 2006). There are interesting advances in this field with the development of various techniques in domains like synthetic chemistry, biotechnology, nanotechnology and gene synthesis. However, these are far from constituting a complete, coherent, and effective set of techniques that will allow the construction and synthesis of the large diversity of machinery and functions that were identified in the preceding text as the "portrait" of the artificial SSR.

- **Micro-electro-mechanical systems (MEMS)** is the technology of very small devices; MEMS are also referred to as *micromachines* in Japan or *microsystems technology (MST)* in Europe. MEMS are comprised of components between 1 to 100 micrometers in size with devices generally ranging between 20 micrometers (20 millionths of a meter) to a millimeter (i.e., 0.02 to 1.0 mm)(Lyshevski, 2000). The technology made significant advances with several types of MEMS currently being used in modern equipment. Some examples include accelerometers, MEMS gyroscopes, MEMS microphones, pressure sensors (used in car tire pressure sensors), disposable blood pressure sensors, and micropower devices. This is probably one of the most promising types of technology for implementing small-scale, artificial SSRs. However, there are still significant hurdles, some of which include implementation of mobile elements (mini robots) and the common difficulty of fabricating the machinery that fabricates MEMS and microprocessors.

- **NASA Advanced Automation for Space Missions 1980 Project.** This study, titled "A Self-Reproducing Interstellar Probe" (REPRO), is one of the most realistic explorations of the design of an artificial "macro" self-replicator and is briefly analyzed in its own section below (Freitas Jr., 1980).

8.1   NASA Advanced Automation for Space Missions Project

One of the missions of the NASA Advanced Automation for Space Missions Project is described in chapter 1 of its final report:

> Mission IV—Self-Replicating Lunar Factory—an automated unmanned (or nearly so) manufacturing facility consisting of perhaps 100 tons of the proper set of machines, tools, and teleoperated mechanisms to permit both production of useful output and reproduction to make more factories. (Freitas Jr. & Gilbreath, 1982)

Later, in the same chapter, the project of making a self-replicating lunar factory is described in detail:

> The Replicating Systems Concepts Team proposed the design and construction of an automated, multiproduct, remotely controlled or autonomous, and reprogrammable lunar manufacturing facility able to construct duplicates (in addition to productive output) that would be capable of further replication. The team reviewed the extensive theoretical basis for self-reproducing automata and examined the engineering feasibility of replicating systems generally. The mission scenarios presented in chapter 5 include designs that illustrate two distinct approaches—a replication model and a growth model—with representative numerical values for critical subsystem parameters. Possible development and demonstration programs are suggested, the complex issue of closure discussed, and the many applications and implications of replicating systems are considered at length. (Freitas Jr. & Gilbreath, 1982)

Figure 10.1 below is a piece of artwork that was reproduced from the NASA study. The description below the picture describes one of the features as a self-replicating factory: "In the lower left corner, a lunar manufacturing facility rises from the surface of the Moon. Someday, such a factory might replicate itself, or at

Figure 10.1: NASA—The Spirit of Space Missions—created by Rick Guidice

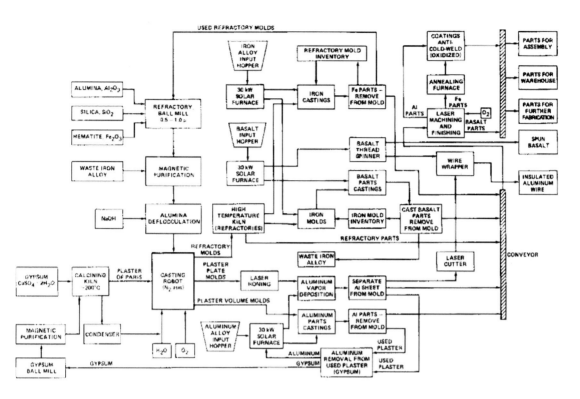

Figure 10.2: LMF Parts Fabrication Sector: Operations (Fig 5.17 in the NASA Study)

least produce most of its own components, so that the number of facilities could grow very rapidly from a single seed" (Freitas Jr. & Gilbreath, 1982).

The project discusses certain details for a self-replicating lunar factory to be feasible. For one, the seed of the lunar factory that would need to be transported from Earth would likely weigh one hundred tons. Additionally, not all of the machinery could be built on the Moon. Certain items, such as computer boards, would need to be brought from Earth since those parts are much too complex to manufacture on the lunar factory. These additional, externally synthesized items are similar to vitamins, which are compounds organisms need to survive but cannot usually synthesize themselves. Scientists estimated that this project would be feasible in the 21st century.

Figure 10.2 illustrates the depth the project reached in considering fabrication facilities on the Moon.

The NASA study represents a thorough, realistic evaluation of the extent that designing a macro-scale (kilometers) self-replicator would entail. It addresses various problems that need to be solved and difficulties that would need to be overcome in order to successfully make a large self-replicator.

8.2   The Self-Reproducing Interstellar Probe (REPRO) Study

In 1980 Robert A. Freitas published a study entitled "A Self-Reproducing Interstellar Probe" (referred to as the REPRO study) in the *Journal of the British Interplanetary Society*. Some of the main goals of the project are summarized below.

- REPRO was designed to be a mammoth self-reproducing spacecraft to be built in orbit around a gas giant such as Jupiter.

- REPRO was a vast and ambitious project, equipped with numerous smaller probes for planetary exploration, but its key purpose was to reproduce. Each REPRO probe would create an automated factory that would build a new probe every 500 years. Probe by probe, star by star, the galaxy would be explored.

- The total fueled mass of REPRO was projected to be 10**10 Kg = 10 **7 tons = 10 million tons for a probe mass of 100,000 tons.

- It would take 500 years for REPRO to create a replica of itself in the environment of a far-away planet.

- The estimated exploration time of the galaxy was 1–10 million years.

9   Simplifying Assumptions for the Design and Construction of an SSR

While the discussion so far has focused on building a fully-autonomous SSR and the difficulties encountered therein, it is possible to reduce the complexity of its design and construction by making some simplifying assumptions. These might include the following:

1. Eliminate the requirement that the SSR produces its own energy. The electrical energy (at an appropriate voltage/amperage) will be supplied to the artificial SSR from the environment.

2. Eliminate the requirement that the SSR has the ability to select, identify and accept through its input gateway appropriate raw materials. All raw materials will be supplied as stock materials to the artificial SSR. An additional, optional simplification could be that all stock materials are labeled appropriately (e.g., with bar codes or RFIDs labels). However, as an illustration, the SSR will still need to use stock copper fed through the input gateways to fabricate copper wires of certain gauges or to use copper in the fabrication of electrical motor parts.

3. Eliminate the requirement that the SSR fabricate the most technically demanding parts, components, and assemblies (e.g., computer boards, microprocessors, semiconductor chips, memories, etc.). These high-technology parts/components (referred to as "vitamins" in the self-replication literature) will be supplied and carefully labeled from the environment through the input gateways.

4. Eliminate the requirement that the SSR must fabricate any part, component, assembly or machinery from basic materials. The SSR will be supplied with pre-manufactured parts that are used in the composition of all its machinery, scaffolding and enclosure. This simplifying assumption means that now the SSR needs to be designed as a (sophisticated) self-assembler that achieves self-replication by assembling exact copies of itself using an exhaustive pool of all of the parts from the machinery/assemblies it is composed of and are supplied by the elementary parts coming through its input gateways.

5. Eliminate the requirement that the information repositories (information catalogs) that drive the functions of the SSR reside within the SSR. This requirement needs to be replaced with requirements for the SSR to possess reliable, high speed communication capabilities to access the information catalogs (and possibly part of the software) residing somewhere outside the SSR. This assumption may simplify certain elements of the SSR design but will make other requirements, such as communication and availability, more stringent for both the SSR and for the external information resource.

Even if the original requirements for the design and construction of an autonomous, artificial SSR are relaxed and any or a combination of the above simplifying assumptions are used as starting conditions for such a project, there are still significant hurdles that need to be overcome in designing and constructing an artificial SSR.

10 Conclusion

As is evident, the three closure rules which must be satisfied by a true self-replicator—energy closure, material closure, and the information closure—place an extraordinary burden onto the design and implementation of self-replicating objects. The SSR must be able to produce and distribute energy, ingest raw materials to fabricate parts, and contain a complete technical description of itself, its processes, and its assembly instructions. Additionally, the environment for the SSR must also be considered, and perhaps specially designed, in order to make available all of the necessary raw materials for self-replication.

While such replicators are known to exist on a biomolecular scale within nature, current technology does not allow for creating a self-replicator at such small scales. The macro scale implementation is more suited to present technology, though this brings its own problems regarding the scale of energy and materials consumption.

While there have been many attempts to build self-replicators, none of them have satisfied all three closure requirements. The NASA REPRO study was the most exhaustive attempt to design a self-replicator, which estimated the replicator to weigh 100,000 tons (unfueled) and reproduce every 500 years.

While Part 1 and Part 2 of this paper contained an overview of the minimal technological requirements of self-replication and foundational technological implementation considerations, the existence of self-replication in nature comes as quite a surprise. Therefore, Part 3 will cover speculative ideas for what the existence of the self-replication process in nature indicates about the nature of reality.

References

Bowyer, A. (2007). The self-replicating rapid prototyper—manufacturing for the masses. In *Proceedings of the 8th national conference on rapid design, prototyping, and manufacturing*: Rapid Prototyping and Manufacturing Association. Available from http://reprap.org/wiki/PhilosophyPage

Drexler, K. E. (1986). *Engines of creation: The coming era of nanotechnology.* New York: Anchor Press / Doubleday.

Freitas Jr., R. A. (1980). A self-reproducing interstellar probe. *Journal of the British Interplanetary Society*, 33, 251–264. Available from http://www.rfreitas.com/Astro/ReproJBISJuly1980.htm

Freitas Jr., R. A. & Gilbreath, W. P., Eds. (1982). *Advanced automation for space missions.* The National Aeronautics and Space Administration and the American Society for Engineering Education. NASA Conference Publication 2255. Available from http://www.islandone.org/MMSG/aasm/AASMIndex.html

Freitas Jr., R. A. & Merkle, R. (2004). *Kinematic self-replicating machines.* Georgetown, TX: Landes Bioscience. Available from http://www.molecularassembler.com/KSRM/5.6.htm

Gibson, D. G. et al. (2010). Creation of a bacterial cell controlled by a chemically synthesized genome. *Science*, 329, 52–56.

Heinemann, M. & Panke, S. (2006). Synthetic biology—putting engineering into biology. *Bioinformatics*, 22, 2790–2799. Available from http://bioinformatics.oxfordjournals.org/content/22/22/2790.full

Lyshevski, S. E. (2000). *Nano- and microelectromechanical systems: Fundamentals of nano- and microengineering.* New York: CRC Press.

Moses, M. (2001). A physical prototype of a self-replicating universal constructor. Master's thesis, Department of Mechanical Engineering, University of New Mexico.

Smalley, R. E. (2001). Of chemistry, love, and nanobots. *Scientific American*, 285(3), 76–77. Available from http://cohesion.rice.edu/naturalsciences/smalley/emplibrary/sa285-76.pdf

Smith, M. (2010). Scientists create first 'synthetic' cells. *ABC News*. Available from http://abcnews.go.com/Health/Wellness/scientists-create-synthetic-cells/story?id=10708502

Stevens, W. M. (2009). *Self-replication construction and computation.* PhD thesis, University of Kent. Available from http://www.srm.org.uk/thesis/WillStevens-Thesis.pdf

Zykov, V., Mytilinaios, E., Adams, B., & Lipson, H. (2005). Self-reproducing machines. *Nature*, 435, 163–164. Available from http://creativemachines.cornell.edu/papers/Nature05_Zykov.pdf

11

Developing Insights into the Design of the Simplest Self-Replicator and Its Complexity: Part 3—The Metaphysics of an Artificial SSR and the Origin of Life Problem

ARMINIUS MIGNEA

The Lone Pine Software

Abstract

This is the last in a three-part series investigating the internals of the simplest possible self-replicator (SSR). Part 1 and Part 2 investigated the necessary design and possible physical implementation of such a self-replicator. This last installment compares potential man-made self-replication to the existing natural self-replicators on Earth, present in the structured hierarchy of ecosystems throughout the world. The insights offered by this series are used to reflect upon possible scenarios for the origin of life and their implications.

1 The Insights into the Design of the SSR, Its Complexity, and the Origin of Life

By way of review, the following are some conclusions that can be made from Part 1 and Part 2 of this study of the design of the Simplest Self-Replicator (SSR), the complexity of the SSR, and the feasibility of construction of an artificial, autonomous SSR.

1.1   The SSR has an overly complex design

The analyses from Part 1 and Part 2 demonstrated that even the Simplest Self-Replicator (SSR) has a strikingly complex design. Beginning with the goal to design an SSR that can accurately reproduce itself and using logical, empirical, and systematic methods, it was shown that the artificial SSR must have a rich set of fully integrated advanced capabilities (functions). Any project or attempt to construct an artificial SSR requires the employment of the most advanced engineering techniques using the latest technologies.

1.2   There are many unknowns about the cell and its mechanisms

Scientists working in molecular biology, genetics, biotechnology, bioinformatics, and related disciplines have made significant progress in understanding the mechanisms of the cell and the information that drives some of its activities. However, it is in the author's estimation that scientists and engineers are still at the beginning of a lengthy road to discover many of the remaining mechanisms, information repositories and processes in living cells and organisms. Here are some areas that are so far still hidden (at least partially) from human knowledge:

- What are the mechanisms for information communication in the cell? Some have been identified, but many still remain to be discovered.

- Where in the cell is the information stored for building the the cell's body plan, including the type, number, and orientation of the cell's organelles? Additionally, will these organelles be linked to each other and other cellular structures? Finally, what is the specific, biochemical composition of each type of organelle?

- How is the "supply-chain" function achieved in a cell, supplying the needed organic material building blocks for protein and organelle construction at the right time?

- What is the nature of the computations performed within the cell? Is it based on protein/enzyme interactions only? Are there any other forms of computations?

- What are the inner mechanisms or control centers that drive cell growth and cell division?

Many of the necessary functions for self-replication elucidated in Part 1 of this study have not been identified within the cell yet, or have only been understood superficially.

1.3   The material basis for an artificial SSR

The most recommended approach for the design and construction of an artificial SSR is based on biochemical materials and a cell-like construction scale. However, the lack of knowledge of so many cellular functions and the technological limits to operating at the cellular level of scale makes this approach impractical and condemned to failure. The goal is to construct an artificial SSR from inert chemicals using a biological material basis, with full understanding and control of all elements and mechanisms involved in such a construction. An accomplishment like Craig Venter's synthetic bacterial cell, although remarkable, is not at all at the level of achievement this would imply, since Venter's project utilized an existing, living organism as the scaffolding for his project.

The alternative approach for building an artificial SSR is to take a macro-scale approach using a common manufacturing/engineering material basis. Compared to what is evidently possible on a biochemical basis, even the smallest miniaturized devices such as smart phones or miniature motors are clunky by comparison. In the analyses in Part 1 and Part 2 of what is involved in the design and construction of even a clunky artificial SSR, design requirements such as the energy closure, material closure, and information closure make the author skeptical that even the most advanced labs in the world would be able to design and construct a fully-functional SSR on a macro-scale.

1.4   Summary of our findings

1. There are many single-celled living organisms that *are fully autonomous* and have a *genuine ability to self-replicate* achieving the energy closure, material closure, and information closure requirements for self-replication.

2. Scientists are still at the beginning of the process of fully understanding the design, information architecture, and biochemical mechanisms of the living world, including the simplest self-replicating single-celled organisms.

3. It can be reasonably estimated that scientists and engineers are not able with the current knowledge and technology to create from scratch an artificial SSR with a biological material basis both because of the lack of understanding of current single-celled replicating organisms and because of the lack of current investigative, operational, and constructive methods for manipulating and fabricating biological-scale artifacts.

4. It can reasonably be estimated that scientists and engineers will encounter enormous difficulties in the design and construction of a macro-scale autonomous SSR with a non-biological material basis because of the difficulty in satisfying the closure conditions at a reasonable scale. No engineering artifact with a complexity approaching that of the investigated SSR, with its level of autonomy and complete automation, has ever been constructed.

5. The prior analyses revealed that the SSR's ability to self-replicate is founded on a full assortment of *highly structured* information resident in the SSR and carried over accurately to each descendent SSR. The information stored in the SSR is highly structured for the following reasons:

   (a) Each *abstract concept* used by the SSR's design (raw material, part, procedure, construction plan, etc.) is represented by a *catalog of entries*, each entry describing an instantiation of that (abstract) category.

   (b) Each entry in a particular catalog (i.e., for a particular abstraction) has a well-defined *set of properties* describing that type of object (entry).

   (c) There are many complex relationships between entries of different catalogs, i.e., between represented abstractions. For example, a part can be made from a particular material; a construction plan is made from a sequence of procedures.

The functional model of the SSR developed here has shown that the SSR must be composed from a large number of well-defined capabilities and mechanisms whose behavior must be integrated, synchronized, and coordinated in their interactions. An SSR cannot grow and duplicate if all these functions are not in place and fully functional. The SSR growth and replication processes cannot be achieved with only one or a subset of the required functions. All must be in place from the beginning. For example, it is not enough to have a fabrication function (RNA) if it is missing the fabrication plan catalog (DNA). Or, even if it has both the fabrication function and the fabrication plan catalog (RNA + DNA) but is missing the input flow control function (membrane controlled pores) and the division control function, the "primitive SSR" will not be able to replicate.

The level of sophistication, autonomy, self-sufficiency, and complexity of the simplest single-celled organisms is beyond the level of technological sophistication achieved by humans thus far. The laws of nature cannot generate highly structured information, because the kind of information needed for the catalogs that are also needed by the SSR are not repetitive or law-like. Likewise, chance cannot be responsible for such information because the catalogs must be finely tuned to deliver accurate information, with small deviations leading to the possibility of catastrophic consequences. Given the rational structure and plan of the cell's internal arrangement, its information, and its sophisticated mechanisms—whose sophistication has only been partially understood by humans because of its enormous complexity—could not be the result of natural processes—the laws of nature or random events.

The belief that the laws of nature and any sequence of natural events and circumstances could have created a self-replicating cell does note have a rational foundation. This author contends that it is purely a statement of faith without a defensible scientific or empirical basis. These considerations lead the author to the conclusion that a naturalistic explanation of the origin of life is impossible. It is unreasonable to conclude that while the smartest human scientists and engineers are currently unable to design and construct an artificial SSR from scratch because of its supreme complexity, some random sequence of natural events could have produced such a sophisticated self-replicator.

2 From the Physics to the Metaphysics of the SSRs

Two hundred years ago the human understanding of both technology and biology was in its infancy. Even then, however, William Paley gave this metaphor about life based only on the biological mechanisms which were known at the time:

> SUPPOSE, in the next place, that the person who found the watch, should, after some time, discover that, in addition to all the properties which he had hitherto observed in it, it possessed the unexpected property of producing, in the course of its movement, another watch like itself (the thing is conceivable); that it contained within it a mechanism, a system of parts, a mould for instance, or a complex adjustment of lathes, files, and other tools, evidently and separately calculated for this purpose; let us inquire, what effect ought such a discovery to have upon his former conclusion.
>
> The first effect would be to increase his admiration of the contrivance, and his conviction of the consummate skill of the contriver. Whether he regarded the object of the contrivance, the distinct apparatus, the intricate, yet in many parts intelligible mechanism, by which it was carried on, he would perceive, in this new observation, nothing but an additional reason for doing what he had already done,—for referring the construction of the watch to design, and to supreme art.
>
> William Paley, *Natural Theology: or, Evidences of the Existence and Attributes of the Deity.* Beginning of chapter II. *State of Argument Continued*, 1809

These analyses have shown that there are strong reasons to be skeptical that humans are able, at this time, to design and build a fully autonomous self-replicator.

Furthermore, there are stronger reasons to be skeptical that the self-replicators found on Earth are the results of natural laws combined with random natural events or circumstances.

Scientists and engineers at NASA made detailed plans and projects to create artificial self-replicators to be realized either as self-replicating moon factories or as self-replicating inter-stellar probes. The time horizon for the implementation of these projects is either sometime during the 21st century or in the distant future. These ambitious projects emphasize the technical hurdles that seem impossible to solve with current technology.

On the other hand, there is a vast assortment of self-replicators populating the planet Earth. The estimated number of organisms on Earth is between 10^{20} and 10^{30}, with an estimated 9.7 million varieties of organisms on Earth. Among those varieties are bacteria, microbes, fungi, plants, algae, grass, shrubs, trees, insects, mollusks, fish, birds, and mammals. Some live in the seas—in lakes, in rivers, and miles below the ocean's surface. Other organisms live in ice, in the Earth's crust, or on the Earth's surface. Some smaller organisms live in other organisms in symbiotic and parasitic relationships. All of these organisms constitute varying positions in the Earth's ecosystem, many providing energy for other biological SSRs. Ocean plankton serves as food for smaller fish and other ocean creatures that, in turn, serve as food for larger fish or ocean mammals. On Earth, grass and plants make up the food for rodents, animals and birds. It appears that for each type of a self-replicator there is a particular food niche suited for it. The three closure requirements–energy, material, and information—which are so difficult to achieve for an artificial SSR are routinely satisfied by the internal design and construction of all of these organisms.

Many of these organisms are not simple self-replicators. Many are significantly more complex than the artificial SSR that was proposed in Part 1 and Part 2. They are comprised of multiple cells and cell types. Additionally, while cell replication does occur, the organism, which is a structured hierarchy of systems of cells, tissues, and organs, replicates in a more sophisticated way at the whole organism level. This study did not consider the SSR's mobility; however, most self-replicating organisms are mobile, which significantly facilitates their ability to feed and replicate. Finally, many self-replicating organisms are endowed with a wide spectrum of sensory organs that allows them to sense the environment.

How can one make sense of the presence of this plethora of autonomous self-replicators? How can one explain the existence of these self-replicators in tandem with a rational, hierarchical structuring of their food chain and a harmonious integration into the Earth's environmental conditions?

This study has pointed to the immense amount of design and coordination *required* to produce even the simplest self-replicator. This means that, at least from the origin of biological life, the Earth has had amazingly intricate machines. As the explanation of the material closure requirement shows, the environment for life to exist must contain suitable materials in a suitable state for use. As the discussion of the information closure requirement reveals, the information to perform self-replication had to exist from the very beginning. As the exposition of the energy closure requirement discloses, each piece of the working self-replicator must be built to be tied in to the power distribution system. This indicates that, rather than being built up over time, life has been infused with design from the beginning. The fact that these intricate interconnections go beyond individual organisms and extend to entire ecosystems indicates that the ecosystem itself, with self-replicators ranging from the smallest single-celled organisms to such multicellular marvels as birds and mammals, operates within the original plan.

Homo sapiens occupies a unique place in this continuum of creatures, being not only a product of the original design, but also a designer himself, with a mind tuned to building things like houses, roads, bridges, engines, cars, airplanes, and planetary exploration vehicles. The mind of *Homo sapiens* looks in amazement to the Earth, at the Sun and planets, and at the galaxy and beyond—it is a mind that dreams of conquering the galaxy. This mind allows *Homo sapiens* to not only operate within the grand design, but also to examine and reflect on it himself, and become a builder and designer within the greater design. With this mind he explores the plants, insects, birds, and animals in the environment, and studies their make and behavior. Further, this mind allows him to look into how these living organisms are constructed, investigating their inner workings. With this mind *Homo sapiens* has learned and is learning about the amazing complexity of living organisms, and this author suggests that he should remain in awe of the magnificent power that created the design at the beginning.

About the Authors

Editors and Organizers

Dominic Halsmer

Dominic Halsmer's love for science and engineering began from the time he was a child playing at the small airport his father and uncles owned. He and his siblings would work on projects to defy gravity, once even successfully building a kite large enough to lift a person off the ground using an old parachute and rope. This natural curiosity eventually blossomed into the work he does today.

Dominic went on to earn his B.S. and M.S. degrees in Aeronautical and Astronautical Engineering from Purdue University, and his Ph.D. in Mechanical Engineering from UCLA in 1992. College was a time of spiritual searching, but ultimately he returned to his faith. "The love and Christ-like example of my parents made the difference," he said. He also notes that the evidence from science for the truth of the Christian worldview also helped draw him back to his faith. Today, he has a heart for reaching out with the gospel to skeptical scientists and engineers.

Dominic joined the Engineering and Physics Department at Oral Roberts University shortly after graduating from UCLA. His interest in flight continues, including participation in the NASA/ASEE Summer Faculty Fellowship Program at NASA Goddard Space Flight Center, and working with undergraduates to test the stability of spinning aircraft under thrust. His current research focuses on studying how the universe is engineered to reveal the glory of God and accomplish His purposes. In 2013 he earned his M.A. degree in Biblical Literature from Oral Roberts University. Dominic is married to his high school sweetheart Kate, and they are the parents of four children. From 2007 to 2012, he served as Dean of the College of Science and Engineering at ORU, and he now directs the ORU Center for Faith and Learning.

Mark R. Hall

Mark Hall is the Dean of the College of Arts and Cultural Studies at Oral Roberts University, where he currently teaches courses on the intersection of science and the humanities, including *Science and the Imagination* and *C. S. Lewis and the Inklings*, as well as other courses in literature. He was recently co-organizer of the "When Worlds Collide" conference on science and science fiction featuring plenary speakers Paul Davies and Joan Slonczewski.

Mark Hall's extensive list of degrees include a B.S.E. in English Education, an M.S.E. in English, and a Specialist degree in Higher Education with an emphasis in English, all from the University of Central Missouri. He has also completed three Masters' degrees from Oral Roberts University including an M.A. in Biblical Literature, an M.A. in Theological and Historical Studies, and an M.A. in Biblical Literature (Advanced Languages concentration). Mark received his Ph.D. in English from the University of Tulsa.

Mark is an ordained minister and a former church pastor. He has been married for over 22 years to his wife, Rachel, who is the Music Director at Jenks United Methodist Church. Dr. Hall and his wife have two children, Jonathan and Kathryne. Mark is active in his community through politics and the theater and is an avid reader, with one of his favorite authors being C. S. Lewis.

Jonathan Bartlett

Jonathan Bartlett is the Director of The Blyth Institute in Tulsa, Oklahoma. The Blyth Institute is a non-profit research and education organization focusing on pioneering non-reductionistic approaches to biology. Jonathan's research focuses on the origin of novelty—both the origin of biological novelty in adaptation as well as the origin of insight in the human creative process.

Jonathan's other roles include managing a team of software developers as the Director of Technology at New Medio, tutoring homeschool students in chemistry and calculus at Classical Conversations, and being part of the Classical Conversations Writer's Circle, where he has a monthly column discussing issues of science, faith, and education. Jonathan is the author of the book *Programming from the Ground Up*, which has been used in Universities from Princeton University to Oklahoma State for teaching undergraduate assembly language.

Jonathan received a B.S. in computer science and a B.A. in religion from Oklahoma Baptist University, and an M.T.S. from Phillips Theological Seminary.

Jonathan and his wife, Christa, have had five boys—three living and two deceased. Jonathan spends his free time practicing Taekwondo with his boys, tending to his garden, and exploring bookstores with his wife.

Primary Authors

Alexander Sich

Alexander Sich is Associate Professor of Physics at Franciscan University of Steubenville. He holds a B.S. in Nuclear Engineering from Rensselaer Polytechnic Institute (with a minor in physics), an M.A. in Soviet Studies from Harvard University, an M.A. in Philosophy from Holy Apostles College & Seminary, and a Ph.D. in Nuclear Engineering from MIT. Professor Sich conducted his Ph.D. research in the Ukraine at the Chernobyl site and worked in the former Soviet Union in the nuclear safety and nonproliferation effort regarding weapons of mass destruction for over thirteen years. In addition to technical articles, Professor Sich has published opinion pieces on nuclear safety (including the Iranian Bushehr issue) in *The Bulletin of the Atomic Scientists*, *The Boston Globe*, The *Wall Street Journal*, *The Diplomat*, and *Newsday*. Professor Sich is married with seven children, speaks near native fluent Ukrainian and fluent Russian. His current research interests lie primarily in the philosophy of nature and in teaching.

Arminius Mignea

Arminius Mignea received his Engineer Diploma from the Polytechnic University of Bucharest, School of Computers. He worked as a software engineer and researcher at the Institute for Computing Technique for 14 years. Employed in the operating systems laboratory, he wrote system software in macro assembly language, Pascal, and C, and published some 12 research papers alone or as part of the team. After immigrating to the United States and settling in California's Silicon Valley, he worked for numerous technology startups developing both server and front-end software for network, enterprise, and security management products. More recently he was involved in the specification, architecture design, and development of software systems for enterprise network traffic monitoring, software build management, and TV server systems. About ten years ago Arminius started being interested in the designs, architectures, and machinery of the biological systems. Arminius is fascinated with the beauty, intricacies, coordination, and incredible levels of organization in living things, and marvels at the skills of their architect.

Eric Holloway

Mr. Holloway is currently an officer in the Air Force and has been on active duty for eight years (including a deployment to Afghanistan in 2010). He has a B.S. from Biola and an M.S. from the Air Force Institute of Technology, both in computer science. The M.S. was focused on Artificial Intelligence with an emphasis on evolutionary algorithms and was funded by an $80k grant from the Air Force Research Lab. The work was published in two conference proceedings, IEEE and ACM.

Winston Ewert

Winston Ewert hails from Canada where he earned a Bachelor's degree in Computer Science from Trinity Western University. He continued his graduate career at Baylor University where he earned a master's degree and is currently a Ph.D. candidate. At Baylor he is currently working for the Evolutionary Informatics Lab, with Robert J. Marks II and William Dembski, dedicated to understanding the role of information in evolution. He has a number of publications in the areas of search, conservation of information, artificial life, swarm intelligence, and evolutionary modelling.

Additional Authors

Tyler Todd

Tyler Todd graduated from Oral Roberts University in 2009 with a degree in Engineering Physics. He currently works for Valmont Industries as a Manufacturing Engineer.

Nate Roman

Nate Roman graduated from Oral Roberts University in 2009 with a B.S. in Engineering Physics. He currently works for Boeing as an Electrophysicist.

Robert J. Marks II

Robert J. Marks II is currently the Distinguished Professor of Electrical and Computer Engineering at Baylor University, Waco, TX. He is a Fellow of both IEEE and the Optical Society of America. He served for 17 years as the Faculty Advisor with the Campus Crusade for Christ, University of Washington chapter. His consulting activities include Microsoft Corporation, Pacific Gas & Electric, and Boeing Computer Services. Eleven of his papers have been republished in collections of seminal works. He is the author of *Introduction to Shannon Sampling and Interpolation Theory* (Springer-Verlag), *Handbook of Fourier Analysis and Its Applications* (Oxford University Press) and is a co-author of *Neural Smithing* (MIT Press). His research has been funded by organizations such as the National Science Foundation, General Electric, Southern California Edison, Electric Power Research Institute, the Air Force Office of Scientific Research, the Office of Naval Research, the Whitaker Foundation, Boeing Defense, the National Institutes of Health, The Jet Propulsion Laboratory, the Army Research Office, and the National Aeronautics and Space Administration (NASA). His Erdős-Bacon number is five.

Rachelle Gewecke

Rachelle was born and raised in rural Iowa and completed her undergraduate studies in psychology at Oral Roberts University (ORU). During her time at ORU, Rachelle participated in a research assistantship with Dr. Dominic Halsmer, an opportunity that produced conference presentations and ultimately a paper examining the concept of reverse engineering nature to discover intelligent design. Rachelle currently resides in Spirit Lake, IA, with her husband, Michael, and daughter, Emily.

Michael Gewecke

Michael Gewecke earned a Bachelor of Arts at Oral Roberts University, a private liberal arts college in Tulsa, OK where he studied Theological/Historical Studies. Most recently he received a Master of Divinity degree from Princeton Theological Seminary in Princeton, NJ. In addition to his academic theological study, Michael has been a technology consultant and online website designer for the last five years. Most recently, Michael served as the CEO of Worship Times, a website design and hosting company that is dedicated to serving small non-profit religious organizations across the United States. These integrative experiences of theology and technology lend towards Michael's interest in the cross section between science and religion with particular emphasis upon their mutuality.

Jessica Fitzgerald

Jessica Fitzgerald is a student at Oral Roberts University pursuing a Bachelor of Science Degree in Engineering Physics with a minor in Mathematics. She plans to attend graduate school for physics or engineering, with a particular interest in nanoscience. Her strengths include data analysis, computation, and attention to detail.

William A. Dembski

William A. Dembski received the B.A. degree in psychology, the M.S. degree in statistics, the Ph.D. degree in philosophy, and the Ph.D. degree in mathematics in 1988 from the University of Chicago, Chicago, IL, and the M.Div. degree from Princeton Theological Seminary, Princeton, NJ, in 1996. He was an Associate Research Professor with the Conceptual Foundations of Science, Baylor University, Waco, TX, where he also headed the first Intelligent Design think-tank at a major research university: The Michael Polanyi Center. He was the Carl F. H. Henry Professor in theology and science with The Southern Baptist Theological Seminary, Louisville, KY, where he founded its Center for Theology and Science. He has taught at Northwestern University, Evanston, IL; the University of Notre Dame, Notre Dame, IN; and the University of

Dallas, Irving, TX. He has done postdoctoral work in mathematics with the Massachusetts Institute of Technology, Cambridge, in physics with the University of Chicago, and in computer science with Princeton University, Princeton, NJ. He is currently a Research Professor in philosophy with the Department of Philosophy, Southwestern Baptist Theological Seminary, Fort Worth, TX. He is currently also a Senior Fellow with the Center for Science and Culture, Discovery Institute, Seattle, WA. He has held National Science Foundation graduate and postdoctoral fellowships. He has published articles in mathematics, philosophy, and theology journals and is the author/editor of more than a dozen books.

Index

CPSIA information can be obtained at www.ICGtesting.com
Printed in the USA
LVOW09*1208280314

379360LV00006B/16/P

9 780975 283868